Introduction to
THERMODYNAMICS

EXPERIMENTAL BOTANY

An International Series of Monographs

CONSULTING EDITORS

J. F. Sutcliffe

Department of Botany, King's College, University of London, England

AND

P. Mahlberg

Department of Biology, University of Pittsburgh, Pennsylvania, U.S.A.

Volume 1. D. C. SPANNER, Introduction to Thermodynamics. 1964

Introduction to
THERMODYNAMICS

D. C. SPANNER

Department of Botany, Bedford College
University of London, England

ACADEMIC PRESS · 1964
London and New York

ACADEMIC PRESS INC. (LONDON) LTD
Berkeley Square House
Berkeley Square
London, W.1.

U.S. Edition published by
ACADEMIC PRESS INC.
111 Fifth Avenue
New York, New York 10003

Copyright © 1964 by Academic Press Inc. (London) Ltd
Second revised printing 1967

All Rights Reserved

No part of this book may be reproduced in any form by photostat, microfilm,
or any other means, without written permission from the publishers

Library of Congress Catalog Card Number: 64-21681

PRINTED IN GREAT BRITAIN AT
THE ABERDEEN UNIVERSITY PRESS LIMITED
SCOTLAND

Preface

This book has been written with a limited objective: to help students who find such a subject as Thermodynamics especially difficult, to acquire a sound grasp of its fundamentals. In view of the fact that no equivalent book seems to be available, and that many discussions of thermodynamic principles in biological textbooks are sketchy, misleading or obscure, no pains have been spared to make this one thorough (within its limits), sound and easy-to-grasp. In pursuance of these ideals the discussions of difficult points have been made much longer and often much more colloquial than usual, and simple analogies have been freely introduced. Footnotes have often been added to lessen the likelihood of misunderstanding or perplexity. The result is that the treatment is often neither rigorous nor conventional; but it is hoped that it is always fundamentally sound and free from misleading implications or suggestions. The aim has been to instil into the student nothing that he will have to unlearn later, but to leave him in a position to appreciate at once the simpler applications of thermodynamics to his own subject, or to proceed readily to more advanced textbooks if he wishes to go further.

Although there is a bias towards problems of botanical interest it is hoped that the book will prove of value to many whose interest lies in other disciplines, such as zoology, medicine, and even chemistry and physics. The earlier chapters are in fact of quite general interest; and even such specialized chapters as that on Photosynthesis (Chapter 14) contain much that is of wide application (in this case to problems of the photoelectric conversion of energy). The same applies to other 'botanical' chapters.

The student will probably find that the treatment is such that the book is better read consecutively than merely consulted; it is certainly intended to be anything rather than a mere compendium of results.

It goes without saying that little in the work is original, and it is a pleasure to acknowledge the help received from the works of previous authors. Principal among these is E. A. Guggenheim whose 'Thermodynamics' is a model of analytical clarity and rigour. D. H. Everett's little 'Introduction to the Study of Chemical Thermodynamics' has also been most useful, and for the last chapter S. R. de Groot's 'Thermodynamics of Irreversible Processes'. These, however, by no means exhaust the list of authorities frequently consulted.

It is a pleasure also to record my thanks to colleagues who have helped by reading the manuscript and offering constructive suggestions, in some cases saving me from serious blunders. The names of Dr L. E. Bentley and Dr N. E. Hill come very readily to mind. I must also thank the publishers for their consistent helpfulness at every stage of the work, Messrs Methuen and E. P. Dutton* for permitting the use of quotations from A. A. Milne, and the Consulting Editors of the series, Prof. J. F. Sutcliffe and Dr P Mahlberg. I would like to add that I would value it if readers who find obscurities, errors or omissions in the treatment would be so kind as to notify me of them, so that should a further edition ever be called for it may be improved in respect of them.

To my wife, to whom I owe a great debt for her understanding and encouragement, this book is gratefully dedicated.

Botany Department
Bedford College
University of London
July 1964

D. C. SPANNER

---

* WINNIE-THE-POOH
From the book WINNIE-THE-POOH by A. A. Milne, by courtesy of Methuen & Co. Ltd, Publishers, and copyright, 1926, by E. P. Dutton & Co., Inc. Renewal, ©, 1954, by A. A. Milne. Reprinted by permission of the publishers.

THE HOUSE AT POOH CORNER
From the book THE HOUSE AT POOH CORNER by A. A. Milne, by courtesy of Methuen & Co. Ltd, Publishers, and copyright, 1928, by E. P. Dutton & Co., Inc. Renewal, ©, 1956, by A. A. Milne. Reprinted by permission of the publishers.

Contents

PREFACE... v
LIST OF SYMBOLS... ix
TABLE OF CONSTANTS.. xii

Chapter

1. The Nature of Thermodynamics............................... 1
2. The First Law of Thermodynamics............................ 9
3. Some Mathematical Topics................................... 21
4. Reversibility and Irreversibility.......................... 41
5. Perfect Gases and Some Other Things........................ 49
6. The Second Law of Thermodynamics........................... 57
7. Entropy and Free Energy.................................... 81
8. Equilibrium and the Direction of Spontaneous Change........ 97
9. The Statistical Interpretation of Equilibrium and Entropy.. 113
10. Chemical Reactions and Membrane Equilibria................ 129
11. Chemical and Transport Processes in Dilute Solutions...... 145
12. Dilute Solutions of Electrolytes.......................... 173
13. The Thermodynamics of Water Relations..................... 203
14. Photosynthesis, Thermodynamic Efficiency and ATP.......... 213
15. The Thermodynamics of Irreversible Processes.............. 233

    SELECTED BIBLIOGRAPHY.. 268

    EPILOGUE... 269

    SUBJECT INDEX.. 273

List of Symbols

Symbols of more General Occurrence

- V volume
- T absolute temperature
- P pressure
- Q quantity of heat
- W quantity of work
- U total internal energy
- H enthalpy or heat function for constant pressure
- F Helmholtz free energy
- G Gibbs free energy
- S entropy
- Δ symbol for 'an increment of' (=final value less initial value)
- d symbol for 'an indefinitely small increment of'
- $\overline{V}_i, \overline{S}_i, \overline{X}_i$ partial molar volume, entropy, etc. of substance or species i
- A affinity of reaction
- n number of moles of substance or species denoted by subscript
- m concentration expressed as molality, i.e. in moles per kilogram of solvent
- c concentration expressed as molarity, i.e. in moles per litre of solution
- M molarities
- x mole fraction
- μ chemical potential
- a activity, relative to value in a standard state
- γ activity coefficient (for use with m)
- γ_{MA} mean activity coefficient for salt MA
- ν stoichiometrical coefficient in a chemical equation
- ξ degree of advancement of a chemical reaction
- K chemical equilibrium constant
- κ partition coefficient of a solute between two immiscible solvents
- ψ electrical potential
- Π osmotic potential

Symbols of more Restricted Use

Chapter 3
- x, y, z, u, w mathematical variables
- F, f symbols for 'some function of'
- p shorthand for $\mathrm{d}y/\mathrm{d}x$

Chapter 6
- C_v heat capacity at constant volume
- C_p heat capacity at constant pressure

Chapter 7
- V_w molar volume of water (i.e. volume per mole)
- $\overline{V}_w$ partial molar volume of water
- p^0 maximum vapour pressure of water
- v^0, v volumes of one mole of water vapour at vapour pressures p^0, p

Chapter 9
- W thermodynamic probability, i.e. number of microstates corresponding to a given macrostate
- N number of counters; Avogadro's number
- m mass
- g acceleration due to gravity
- h height

Chapter 12
- ϵ quantity of electricity
- z valency or charge number of an ion
- λ absolute activity
- $\mathscr{F}$ the Faraday, i.e. total charge carried by a mole of univalent ions (*see* Table of Constants, p. xii)
- I ionic strength of a solution
- ρ accumulation ratio of an ion
- $\mathscr{D}$ relative Donnan strength

Chapter 13
- σ a coefficient (approximately unity) for converting molality to molarity (i.e. $c = \sigma m$)
- $S.P.$ suction potential of a phase ($S.P. = \Pi - P$)
- p vapour pressure of water in equilibrium with phase
- $\overline{V}_w$ potential molar volume of water

Chapter 14
- η efficiency
- λ wavelength
- ν frequency
- I radiation intensity, i.e. energy per unit time falling on unit area
- $\mathscr{I}$ specific radiation intensity, i.e. intensity per unit frequency interval

Ω solid angle
δs element of area
u radiation density, i.e. energy per unit volume
subscripts $_r,\ _l$ indicate 'radiation' and 'leaf' respectively

Chapter 15

t time
a general symbol for something changing with time
L conductance coefficient
J flux
X force
σ rate of production of entropy
v,u,h,g,s values of V,U,H,G,S per mole
subscripts $_{m,\ U,\ Q}$ mass, energy, heat
superscript $*$ denotes a 'quantity of transfer'

SUBSCRIPTS

$_{1,2}$ used in Chapters 2, 6, 7 and 9 for two different states of a closed system

$1, 2, \cdots _i$ used to indicate different chemical species; in particular w indicates water, s an unspecified solvent, and the normal chemical symbol is used where appropriate

$(g),(l),(s)$ substances in gaseous, liquid, and solid states respectively

$_{A,B,C,D}$ belonging to the chemical reactants A,B,C and D (as in $\mu_A, \nu_B, n_C, \gamma_D$)

$_{\text{rev.}}$ indicates that the quantity is measured along a temporal pathway traversed reversibly (e.g. $Q_{\text{rev.}}$)

SUPERSCRIPTS

$^{\alpha,\beta}\cdots$ used to indicate phases or sub-systems

^{0} indicates a standard value for pure substance ($x = 1$)

$^\ominus$ indicates a standard value for solute at unit molality ($m = 1$)

$^{-}$ partial molar quantity (e.g. $\overline{V}_w$); mean value (activity coefficient)

Table of Constants and Units

R gas constant = 1·987 calories per degree per mole
= 0·08206 litre atmosphere per degree per mole
= 8·315 × 10^7 ergs per degree per mole
= 8·315 joules per degree per mole

N Avogadro's number = 6·023 × 10^{23} per mole

$k = R/N$ Boltzmann's constant = 1·3805 × 10^{-16} erg per degree

h Planck's constant = 6·625 × 10^{-27} erg second

$\mathscr{F}$ the Faraday = 96,493 coulombs per gram equivalent

c velocity of light = 2·9979 × 10^{10} centimetres per second

T_{ice} absolute temperature 0°C = 273·15°K

ln (logarithm to base e) = 2·3026 log$_{10}$

1 kilocalorie = 1000 calories
1 calorie = 4·184 joules = 4·184 × 10^7 ergs
1 litre atmosphere = 24·21 calories
1 atmosphere = 1·01325 × 10^6 dynes per square centimetre

The mole is the quantity of substance in grams whose numerical measure is the chemical formula weight. Thus:

one mole of water (H_2O) = 18·00 grams
one mole of sodium ions (Na^+) = 23·00 grams

The mole contains Avogadro's number of the entities to which it refers.

CHAPTER 1

The Nature of Thermodynamics

> One day when Pooh Bear had nothing else to do, he thought he would do something, so he went round to Piglet's house to see what Piglet was doing. . . .
>
> He expected to find Piglet warming his toes in front of his fire, but . . . the more he looked inside the more Piglet wasn't there. 'He's out', said Pooh sadly. 'That's what it is. He's not in. I shall have to go a fast Thinking Walk by myself. Bother!'
>
> THE HOUSE AT POOH CORNER
> *A. A. Milne*

Thermodynamics is the study of the interrelations of matter and energy, these interrelations being expressed with the help of quite simple mathematics in an exact form. Yet to describe it in this way is not enough, for there are several other branches of physical science of which much the same could be said. Mechanics is an example, though the mathematics in this case is rather more elaborate. Thermodynamics is different from the latter in the way in which it approaches and handles its subject matter, and it will be useful if a moment is spent considering what may be called, in this connection, its special attitude to its material.

In the first place thermodynamics studies *systems*. The reader will often come across this term; in fact it is one of the characteristic words of this subject. The dictionary defines a system as a 'complex whole', and that is in fact how thermodynamics views matter. Its systems may undergo 'internal change', they may 'interact', or they may be split up for convenience into sub-systems; but the fact that they are 'complex wholes' is never really lost sight of. By contrast, mechanics studies *bodies*, which possess a simple individuality rather than a complex wholeness. One thinks of weights, pulleys, tie-rods, springs and so on. It is true that mechanics can assemble its simple bodies into mechanical systems; but if this process goes far enough—one recalls the derivation of the properties of gases by the laws of mechanics applied to a vast assemblage of individual molecules—then the results derived become similar in kind to those found by the thermodynamic approach. In other words as the idea of a system—a 'complex whole'—comes to the fore, so do the findings of mechanics often acquire a likeness to those of thermodynamics. Broadly, therefore it can be said that while mechanics studies the external relations of bodies, thermodynamics concerns itself with the internal changes in systems.

In the second place however, thermodynamics limits itself very drastically even in its study of systems; how and why it does this will become apparent from a few examples. Suppose some water is placed in a stoppered bottle; then the vapour pressure above the water at a particular point may have any value. It all depends on the history and geography of the system, that is on such particulars as how long the water has been in the bottle, whether there are internal obstructions to gaseous movement, and so on. However, there is one qualification that removes all uncertainty. If it is specified that the system (in this case the bottle with its contents) has reached internal equilibrium, then it can immediately be said that the vapour pressure inside has a perfectly definite and unique value, which can be tabulated for future reference. Again, suppose a suitable enzyme preparation is added to a mixture of organic substrates. Consider what proportions of the reacting substances are present later. It is apparent that these proportions may have any value; it all depends on how much of the various substances were taken to begin with, how active the enzyme preparation is, and how long the reaction has been in progress. But if equilibrium has been established all uncertainty vanishes. The proportions then fulfil a simple relationship summarized by the equilibrium constant.

A third example is somewhat different. In modern manufacturing processes the development of high vacua in large and complex vessels is often a requirement; and sometimes, understandably, small leaks develop and are very troublesome to locate. At the site of such leaks a small stream of air molecules is passing at high velocity from atmospheric pressure into virtually a perfect vacuum. Suppose observations could be made unhampered by technical difficulties and the motion of each molecule tabulated very soon after it had entered the vacuum vessel. Suppose further that a frequency diagram, or histogram, were to be plotted showing the relative numbers of molecules possessing kinetic energies within given ranges (just as the histogram of beans on a basis of weight might be recorded); should we find any definite and unique pattern in the diagram? The answer is, no; the diagram would depend in a complex and practically indeterminate way on the position at which we chose to do our sampling. Now suppose observations were made far enough away from the leak for the incoming air to have dissipated its headlong motion and to have achieved an internal equilibrium as a body of gas at rest. In this case the pattern of the histogram of molecular energies has reached a quite definite and stable form, very closely related in fact to the normal distribution so familiar to statisticians. Its features could be tabulated and recorded for future reference, just as in the previous examples.

These instances serve to indicate the second fundamental characteristic of thermodynamics. This can be expressed by saying that thermodynamics takes as the means for describing and characterizing its systems those properties for whose measurement internal equilibrium is presupposed. In the two earlier cases quoted—those of the vapour pressure of water and of the equilibrium constant of a chemical reaction—this is obvious enough; but it often escapes notice that such variable parameters as temperature and pressure, and of course entropy, are only really defined for systems in internal equilibrium. That this is so can be illustrated by the third example. Suppose an attempt is made to measure the temperature or pressure of the air stream very close to its point of entry. It would be discovered that the reading of the thermometer or probe used would vary disconcertingly with such things as its orientation, shape or surface roughness, factors which should have no influence. The instrumental indications might be highly instructive; but they could obviously not be the straightforward temperature and pressure of the air. On the other hand once the truly random pattern of molecular Brownian movement has been established the instrumental readings assume quite unique values; shape and orientation have ceased to be relevant and this is as it should be. 'Temperature' and 'pressure' have exact meanings. Of course all this does not imply that the very slightest departure from equilibrium invalidates these concepts. It is a progressive process; and in practice such properties as temperature and pressure may be invoked up to the point at which the instruments begin to show erratic behaviour. As a matter of fact the random pattern mentioned previously (often called the Maxwellian distribution) is established so quickly compared with the equilibria which the biologist or chemist habitually studies, that no difficulty at all is found in speaking of the temperature, pressure or entropy of a system in which these other equilibria are far from obtaining. This is a most fortunate state of affairs, for were it not so the biologist, at least, would be denied altogether the use of thermodynamic variables (T, P, S, etc.) as conceptual tools for describing his systems.

This limitation to properties defined at equilibrium has several important consequences for the subject of thermodynamics. Firstly it introduces a considerable simplification into the mathematics, and enables thermodynamics to deal with situations much more elaborate and complex than could be tackled easily by other lines of approach; this is a very valuable feature. Secondly it means that thermodynamics—at least what is known as classical thermodynamics—has nothing to say about rates. The very notion of a rate implies a system not only out of equilibrium but also in the very process of changing, so this is not surprising. It is true that the more recent development called irreversible thermodynamics does say something about rates; but even here the rate must strictly be measured indefinitely close to equilibrium, and so the general nature of this aspect of thermodynamics is not really affected.

While the notion of a *rate* of change is thus foreign to the classical branch of the subject that of a *direction* of change is not, and for an interesting reason. The former involves properties like viscosity or permeability measured *while the system is changing;* the latter need involve only bona fide thermodynamic properties measured *while the system is prevented from changing*. Thus it might be questioned whether a mixture of starch and phosphoric acid would change spontaneously into hexose phosphate, or whether the spontaneous *direction* would be the reverse one. This can be handled thermodynamically by supposing the reaction to be temporarily 'poisoned' or inhibited (not a very far-fetched requirement) and then all the observations can be made on a system at rest. By comparing these with similar observations on the system 'poisoned' at another point on its course (i.e. with different proportions of all the reactants present) thermodynamics can be used to give an authoritative answer to the question, which is the spontaneous direction in which the system will move if allowed to do so? Where the change is not a chemical one but say of the nature of a transport process, the 'poisoning' can of course be done very easily with an impermeable wall. The important point to remember is that this sort of question—and it is one of extreme importance to the physiologist—is one which can be answered strictly within the limits of the subject, whereas questions of rates cannot. At best, they require a sort of hybrid approach whose validity is open to question.

Systems not in Equilibrium

It will be seen later that there are many different sorts of equilibria (such as chemical, thermal, osmotic and so on) and the question sometimes arises in biological contexts, if thermodynamic results are really only valid for systems in equilibrium is it right to apply them to systems (like living cells) which are known to be out of equilibrium? To take a concrete example, it is known that a living root hair is definitely not a system in equilibrium. It is the site of continuous oxygen uptake, of carbon dioxide production, of protoplasmic streaming and so on. Can its water relations, therefore, be expected to be governed by the classical (and typically thermodynamic) equation:

Suction Potential (or D.P.D.) = Osmotic Potential of cell sap −
Turgor Pressure.

Or to be even more specific, if the terms on the right hand side cancel out, can it be taken that, although the cell is not in *absolute* equilibrium in *every* respect, there will be no water uptake?

Now this is really another way of stating the problem of whether or not there is present what biologists have come to call an active mechanism of water transport and it is a question which cannot be decided *a priori*. It is a fact of observation that an equilibrium of one sort may be apparently quite unaffected by a lack of equilbrium of a different sort in the same system; for instance the ionic equilibrium which gives a steady value to the pH of the cytoplasm may be quite untouched when water begins to flow rapidly through the root cell in obedience to a sudden transpirational pull. On the other hand it is at least conceivable (though perhaps very unlikely) that the transpiration current through the cytoplasm might upset an enzyme-catalysed reaction; large participant molecules like starch might be swept away from the enzyme surface significantly more than smaller ones like glucose phosphate, giving the latter as it were an unfair advantage. The result of this might be that the steady bulk concentrations of the reactants no longer yield the thermodynamic equilibrium constant. Although this particular example is quite hypothetical and probably far-fetched the existence of active mechanisms is well established in living systems, and in fact the failure of thermodynamic equations to hold is one of the best evidences for them, the accumulation of ions by living cells being a case in point. Such instances naturally constitute cases of very great interest to the physiologist. While it is true that thermodynamics gives little insight into their mechanism it at least does this considerable service; it draws attention to linkages between phenomena which otherwise might have been rather unsuspected. More will be said about this later. The answer to the question therefore as to whether thermodynamics can be applied to systems, like living ones, known to be not in total equilibrium is just this: if the lack of equilibrium concerns the very process to which thermodynamics is being applied (say the uptake of water), then of course the equations will be found not to fit the facts; the suction potential equation is not expected to hold if the water uptake is not completed. If, however, the non-equilibrium concerns a fundamentally different process (like the continued utilization of oxygen, or synthesis of protein, in a water relations experiment) then the validity of applying the thermodynamic equation becomes a much more interesting problem. If it is still found to be valid—and this is a matter which can only be decided by experiment—then all is well; but if it no longer holds then it must be recognized that in the experimental organ some other process is exerting an 'active' influence on the process being studied. Thermodynamics may further give some clues as to what other previously unsuspected processes may in fact be interfering in this way; their energetic turnover must be big enough to produce the observed effect, as will be discussed later.

What in fact physiologists do is this. Firstly, they make the presumption that living things, though far more complex than inanimate ones, are just as much subject to thermodynamic laws, a presumption which there is as yet no sound reason to question. Secondly, they make the much lesser presumption that the process which for one reason or another is of immediate interest can be considered in isolation; its position of 'equilibrium', and its urge to reach it, are determined by itself alone. In following out this lesser presumption it may, or may not, be found that it justifies itself. If it does, one can say that the process is determined only by 'passive' forces; if it does not, the conclusion must be that some hitherto unsuspected 'active' mechanisms are interfering. It is possible of course to be much more radical and to call the first presumption into question; but that would appear to be to go into a blind alley. Certainly experience suggests that it is much more fruitful to question the second and lesser presumption. In fact, it seems *a priori* to be clear that every process must have *some* influence on every other process in a living thing, or for that matter in a non-living one· It may be a quite insignificant influence, and so may safely be ignored; but it may be a highly important one, as is apparently the case between respiration and salt uptake, or ionic diffusion and water movement through a membrane. Living things are the site of such an enormous range of different activities that the scope for this 'active' influence of one process on another is potentially enormous; and every time it is noticed that the intact organism is a highly integrated thing this potentiality is emphasized. Therefore as physiologists we have to be continually on the look out for active mechanisms; but their possibility, or even probability, does not mean that thermodynamics ought not to be applied on the grounds that the systems are not in internal equilibrium. It means rather that thermodynamics must be applied.

Thermodynamics and Molecular Theory

There is another characteristic of classical thermodynamics that must be mentioned, and that is its independence of the molecular theory of matter. This can be seen at once from the consideration that as a matter of history the subject was built up from a simple fact of everyday observation, namely that it is impossible to construct a machine to give perpetual motion. Even the planets (enlarging our idea of construction a little) are slowing up. It is on this simple basis, with one or two elementary additions, that the whole structure of classical thermodynamics stands. It is true that there are good reasons for believing that in the atomic world perpetual motion does occur; electrons move unceasingly in their orbits around the nucleus. Occasionally such perpetual motion rears its head into the world of large scale happenings as in the well-known phenomenon of superconductivity at the excessively low temperature of liquid helium. But these apparent exceptions merely serve to emphasize the sphere in which classical thermodynamic laws rule, that of systems large enough to be seen with the naked eye (i.e. macroscopic), and at temperatures not too close to the absolute zero[1]; in fact just where the

---

[1] It is not implied that thermodynamic laws fail near the absolute zero, but only that their validity may then depend on other arguments.

impossibility of perpetual motion is our unvarying experience. Brownian movement, which of course goes on interminably, need not concern us here, for it has no direction associated with it and is consequently not a process. In other words it does not represent the sort of perpetual motion on the impossibility of which the development of the subject is based.

While, however, thermodynamics has been developed on the basis of phenomena which can be observed in systems at least large enough to be handled in the every day sense, the assumption of an underlying molecular constitution of matter does throw a flood of light on the inner meaning of the whole subject. In fact the great thermodynamic generalizations have been derived again on the basis of molecular theory and the quantum laws, with no reference to perpetual motion. The treatment, which constitutes the science of Statistical Thermodynamics, is highly mathematical, and will only be drawn on when an endeavour is made to make the conceptions of entropy and the nature of equilibrium seem a little less difficult to grasp. For the moment it is sufficient to remark that its relationship to Classical Thermodynamics is in some respects rather similar to that of cytology to mathematical genetics. Cytology is based on observable phenomena, like gametes, chromosomes and chiasmata; genetics, more highly mathematical, arises at the fundamental level by postulating the hypothetical entities called genes. Both arrive, by different routes, at an explanation of the practical behaviour of plants and animals in inheritance. The analogy is not a very exact one, but it is instructive.

Thermodynamics and Mechanism

It has already been remarked that Classical Thermodynamics starts off from the very general observation of the impossibility of making a perpetual motion machine, and it was seen that this defines the sphere in which its conclusions are valid, a sphere usually spoken of as the macroscopic one. Now, in so far as the fundamental observation is a perfectly general one, making no provisos as to the sort of phenomena used in designing the machine, thermodynamic results themselves are not tied down to any particular mechanisms. They are perfectly general and apply no less to the nuclear generation of power than to its generation by chemical, osmotic, electrical or any other means. The science of mechanics on the other hand as its name suggests, is very much concerned with mechanism. As a consequence of the fact that thermodynamics is not, it reaches conclusions true for all mechanisms. In avoiding entanglement, as it were, in the fine details of the processes it considers, thermodynamics achieves much more easily and quickly than mechanics a general grasp of the outcome of things, and in so doing it is able to include in its treatment from the start the very obvious and universal property of temperature, a concept which only arises for mechanics in its relatively old age, as it were. The idea of temperature (and heat) is, in fact, the one which more than any other suggests itself as *the* characteristic concept of our subject, at least to the non-specialist. In fact entropy stands on an equal footing, as belonging peculiarly to thermodynamics; but its familiarity to the biologist is hardly quite so great.

The Uses of Thermodynamics

To what sort of results does thermodynamics lead? This is a question that is bound to occur to the reader very early in his acquaintance with the subject. Broadly speaking, it leads to two kinds of results. Firstly, it gives answers to the question discussed earlier: in which direction would a system change spontaneously if all physical barriers were removed? Included also are such chemical barriers as enzymatic poisons or even the absence of enzymes. This is far and away the most important function of thermodynamics to the physiologist. Secondly, it leads to relations between seemingly quite unrelated equilibrium properties of substances and systems; for instance, the e.m.f. of a galvanic cell and the equilibrium constant of the chemical reaction on which it depends, or the osmotic potential of a solution and its vapour pressure. While these relations may not, in themselves, be of interest to the physiologist as such, he will find that they are often an essential stepping stone in his search for answers to his problems. Finally, thermodynamical ways of thinking represent a discipline, and in this they have a value over and above the results to which they lead. The student who has studied the subject successfully has acquired a way of looking at things, a feeling for one of the most fundamental aspects of natural happenings that will often be of value to him when pondering over questions of mechanism, in spite of the fact that thermodynamics itself transcends mechanism and has nothing in detail to say about it. This is therefore a further and weighty incentive or the study of the subject; and one which it is hoped the further reading of this INTRODUCTION will not disappoint.

CHAPTER 2

The First Law of Thermodynamics

'I'm not complaining,' said Eeyore, 'but There It is.' Pooh sat down on a large stone, and tried to think this out. It sounded to him like a riddle, and he was never much good at riddles, being a Bear of Very Little Brain.

WINNIE-THE-POOH
A. A. Milne

WORK

The idea of work is one with which fortunately or unfortunately according to circumstances, everyone is familiar. Leaving aside the conception of mental work, which is something quite different, it is apparent that there is a common element in all the various familiar forms of muscular exertion: digging the garden, carrying packing cases upstairs, or sawing logs. Moreover, it is felt intuitively that these experiences should be expressible quantitatively, for it is obvious that 'more' work is done on one occasion than on another. Scientifically therefore the search for quantitative relationships in nature is bound to lead to the question, how can a precise measure be established of the element that is common to all these experiences, and gives them their distinctive character as work? In looking for such a measure it is natural to start with concepts which are themselves at once intuitively simple and capable of easy measurement; then 'feel the way' with these until some definite progress is made. That is how the idea of mechanical work arose. In all the experiences of work it is found that a muscular force must be exerted (easily measureable by balancing against a weight or by using a calibrated spring) on an object, like a spade, which moves some distance in response. What should then be more natural than taking the product of force and distance as a measure of the work done? When this suggestion is carried out it is found that the product thus defined is indeed a most valuable and useful function; it justifies itself by results. It is important to realize this because conceptions like entropy and free energy to be dealt with later make less immediate appeal to the intuition on the basis of everyday experience than does this idea of work, and the biologist may be left with the uncomfortable feeling that he does not quite know where these ideas have come from, or what their real status is. Are they human inventions, or does nature really possess a stuff called entropy? The fact of the matter is simply that frequently in the quantitative study of a chosen field of natural phenomena the suggestion arises that a certain way of combining the measurements obtained may prove useful, and produce a helpful tool in correlating and describing experiences. When tried, this does indeed lead to very useful and profitable lines of thought. It is in this way that concepts like energy, momentum, work, entropy and so on have arisen. It is analogous to the situation in which, faced with the problem of how to eat, the idea of knives, forks and spoons presents itself. These prove so convenient that they pass into universal currency; in time their status is forgotten, as it were, and they are somehow regarded as belonging to the very structure of nature, instead of as being tools developed by humans and justified by their effectiveness.

The Conservation of Energy

As an illustration of the great fruitfulness of the concept of work the Principle of the Conservation of Energy may be discussed. Having defined work in the way mentioned above, certain other ideas suggest themselves. For instance, if the system under investigation contains a moving body (think for instance of a pendulum) the product $\frac{1}{2}mv^2$ frequently occurs (m being the moving mass and v its velocity); in fact it is found that when a force does work on the body this product ($\frac{1}{2}mv^2$) undergoes a change exactly equal to the work done, both in sign and magnitude. Moreover, when work is done it produces changes other than altering the speed of moving bodies. Springs can be stretched, for instance, and weights raised. In the former case, the product $\frac{1}{2}sx^2$ puts in an appearance (s referring to a property of the spring and x being the distance by which it is stretched); and in the latter case, the product mgh (where m is the mass raised, g the acceleration due to gravity, and h the height). All of these changes can be accomplished by muscular effort, and physiologically we are conscious that this accomplishment uses up our 'energy' and we become tired. This links the idea of energy (physiologically defined) with that of work. Further, these various changes are in an important sense interconvertible; any one can be used in reverse to promote the others and provided friction can be eliminated, the conversion is complete and can go backwards and forwards repeatedly. Thus it is a matter of common experience that a stretched spring can set a body in motion, and the latter in turn can stretch the spring; further that the cycle can be repeated for a number of times limited only by the completeness with which frictional effects have been eliminated.

These ideas arising from physiological sensations and extended into the inanimate world, lead on naturally to the principle under discussion. 'Energy' is needed for work and the doing of work produces changes of different kinds (motion, elastic deformation and so on); these different kinds are interconvertible repeatedly and without loss, provided care is taken to eliminate friction. The changes would therefore seem to be storing 'something' when they move in one direction which they release when they move in the other; something which, again in the absence of friction, is never lost (or the changes could not go on repeatedly). It is only necessary to call this 'something' energy and to recognize that it can exist in different forms[1], to have arrived at the Principle of the Conservation of Energy. Had we started by trying to measure the intuitive concept of work by the product of force and (distance)[1] such a fruitful result would never have been found; the measure adopted has therefore been justified by its results, and it is not surprising that like spoons and forks, it has passed into a very general currency. It will be found later that the same thing is true of our thermodynamic concepts.

The First Law of Thermodynamics

The mechanical Principle of the Conservation of Energy specifically excludes losses due to friction. This is a situation which 'grates', if one may express it like that; one feels that such an elegant and powerful principle ought to be of universal truth, or at any rate that there should be some satisfying reason for its inapplicability in the situation indicated. However, circumstances are not lacking that suggest a way in which this obstacle to universality can be overcome. Wherever frictional forces are called into play and oppose motion there is a tendency, very frequently and easily noticed, for the temperature to rise; it is often said, rather loosely, that heat is developed. If this heat can be regarded as somehow another form of energy, differing in some respect from other forms of energy in a way that puts it outside the Mechanical Law of Conservation, then we can perhaps include it in a more general statement. That heat can be regarded as a form of energy was not always an accepted interpretation; one has only to recall the old 'caloric' theory of heat as an unusual fluid substance to realize this. From the present day position however it seems perfectly reasonable, since temperature is now thought of as reflecting the kinetic energy of the random Brownian movement of the molecules. A rise of temperature does consequently mean that, at a hidden level, energy has become more abundant. Proceeding on these lines therefore, it can be supposed that friction does not destroy energy; it merely causes it to be dissipated in a very disorderly way amongst the molecules, thus raising their individual energies. That this is a justifiable point of view is borne out by experiments in which the amount of heat developed by a given expenditure of work done against frictional forces is carefully measured; the relationship turns out to be always the same (the so-called 'mechanical equivalent of heat', first measured by Joule). As a

---

[1] Measured by the products ($\frac{1}{2}mv^2$, $\frac{1}{2}sx^2$, mgh, etc.) previously mentioned.

result it is now considered to be reasonable to believe that heat is a form of energy, and that as such it is subject to a widened law of conservation. This widened law is what has come to be known as the First Law of Thermodynamics. From the macroscopic point of view (that is, confining attention broadly to systems large enough to be seen by the eye or manipulated by the hand) this extension is a real one, since a rise in temperature is obviously a very different sort of thing from an acceleration in motion or an elastic stretching. From the microscopic point of view (putting ourselves in the position of atoms or molecules) the extension is not so clear cut; a rise in temperature is actually visualized as an acceleration in motion and an increase in the vigour of elastic vibration. However, there is a difference; a special factor which the atoms would notice is that the acquisition of energy in the form of heat increases the sense of chaos among them, whereas its acquisition in a 'mechanical' form does not.

Heat, or thermal energy as it is perhaps better called in the present connection, is not the only new category of which the First Law must take cognizance if it is to be of the widest generality. When a simple gas[1] is slowly compressed in contact with an ordinary thermostat bath the work done in compression is converted into an exactly equal amount of thermal energy, which passes as heat into the bath. On the other hand, when a mixture of reactive gases such as hydrogen, nitrogen and ammonia is similarly compressed (in the presence of a suitable catalyst), the heat produced is no longer equal to the work done, but exceeds it. It is found concomitantly that compression has driven the chemical reaction in the direction of the production of more ammonia, a result which is in keeping with Le Chatelier's Principle. Expansion of the gas mixture to its original volume produces the opposite effects, and restores the *status quo*. Thus, just as one can postulate potential energy associated with elastic strain or height as terms in the mechanical principle of conservation, so now chemical potential energy[2] may be postulated as a category in the thermodynamical First Law. In fact, to extend the parallelism, it is found that when hydrogen and nitrogen react in a rigid closed vessel (so that no work is done) the chemical change results in the evolution of heat, as in the case of (though less noticeably) a tightly-coiled spring which is suddenly released. Thus, when a chemical potential energy term is included it may again be maintained, consistently, that energy has neither been created nor destroyed in the compression of the gas mixture.

It need hardly be added that chemical energy is not the only new form that might be discussed. Nuclear and other forms are embraced too, though none is nearly so important to the biologist.

[1] Strictly, a perfect gas (Chapter 5).
[2] It is the decrease in this when more ammonia is formed that accounts for the extra heat in the second case.

The Meaning of Heat and Work

The foregoing discussion of forms of energy may have left the reader somewhat uncertain. The First Law, it has been implied, embraces all conceivable forms of energy; and work, heat, thermal energy and chemical energy have been referred to among others. What has perhaps not been made plain is that these categories are not all, as it were, logically alike. Forms of energy possessed by a material system can be divided conveniently into two: those in which the energy is held in a dynamic fashion, and those in which it is held in a static one. In the former are ordinary macroscopic kinetic energy and the kinetic part of the thermal energy[1]; in the latter are the potential[2] energy associated with gravity, elastic deformation and chemical configuration. Now comes an important distinction: work and heat do not belong to this classification at all, for they are not forms in which energy is 'held'. They are ideas which characteristically arise when considering *how energy is passed between one material system and another*; thus we may say that system A passes energy work-wise or heat-wise to system B. What actual form of energy system A surrenders when it does work on B or communicates heat to it, is another question altogether; it may draw on its resources of chemical, electrical, elastic or even thermal energy. The same applies to the form in which system B stores the energy it has received. Thus it can be asserted that heat and work are not two forms of energy; they are better regarded as the two modes in which energy is communicated.

Here, however, a warning should be noted. The terms heat and work are in fact used in senses other than those attributed to them up to this point. Some of these senses are worthwhile extensions of the original concepts but others are misleading. As an example of the former the reader may well come across the expressions 'chemical work' or 'osmotic work' in biological contexts where the notion of force times distance is certainly not in mind; in fact in his alternative and very helpful treatment of thermodynamics Brønsted gives the term work from the start a much wider meaning than the one we have used. Brønsted (1955)[3] also gives a slightly different meaning to the term heat. There is little need for the ordinary biological reader to have these different usages at his finger tips, but the point is mentioned because it may cause unnecessary confusion. He will understand from one writer that under suitable conditions heat can be converted into work; he may then discover from Brønsted that under no circumstances is this conversion possible. No contradiction is implied, only a difference in definition.

[1] Thermal energy where molecular vibrations are present also has an elastic (i.e. potential) component.

[2] The reader may like to be reminded that potential energy is energy associated with position, or spatial configuration. This explains the term 'chemical potential energy'.

[3] Brønsted, J. N. (1955). 'Principles and Problems in Energetics'. Interscience Publishers, Inc., New York.

There is, however, a common usage of the word heat which should be noted as liable to mislead. The notion of thermal energy—as the sum total of all those elements of both kinetic and potential energy being continually and rapidly randomized between molecules in the processes of molecular collisions—is a sound one; and thermal energy is sometimes spoken of as heat, for instance when it is said that the energy of a bullet is converted on impact into heat. This sort of usage however is liable to mislead. When a system 'absorbs heat' (this is the correct usage) it is only in simple cases that its thermal energy increases by an exactly equal amount; 'heat' and 'thermal energy' are not really interchangeable ideas. Where chemical processes or phase changes like evaporation occur, the increase in thermal energy may be much less than the heat absorbed; some of the latter may be locked up in a 'latent' or chemical form, for instance, as potential energy. Thus, to re-emphasize, it is preferable to limit the term heat to energy communicated in a particular way to a system; once the system has received it we no longer speak of it as heat. It is held as thermal energy, as potential energy of molecular separation or chemical configuration, or as some combination of these.

Mathematical Formulation of the First Law

In order to use the First Law to develop quantitative results it must be put into an exact form. As the argument used to do this is a rather general one which will be of service later, a little time will be spent in developing it.

We first assume that we are dealing with what is called in thermodynamics a 'closed system'; that is, a fixed quantity of matter. It can be simple or complicated, just a plain metal spring or a whole intact plant or organ, dead or alive. The system is considered in two different 'states'; that is, in two conditions in which it is observably different (using the word 'observably' in a wide sense to cover the readings of any sort of instrument such as a thermometer, a densitometer or a voltmeter. If the instrumental readings are different as between the two states, that is enough.) The states will, at least in thought, have to be specified; and this is done by naming their observable, that is to say measurable, characteristics (this again focuses attention on the fact that thermodynamics deals only with *macroscopic* systems, ones which can be instrumentally handled). There is of course a very wide range of observable characteristics from which to choose, such as pressure, temperature, volume, refractive index, dielectric constant, hardness and magnetic susceptibility, to name only some; but fortunately experience shows that it is not usually necessary to specify more than a few of these in order to define completely the state of the system. In other words, they are not all independent; expressed differently again, the number of degrees of freedom of the system is often quite small (except of course with complex systems like organisms).

2. THE FIRST LAW OF THERMODYNAMICS

It may be that only two or three have to be specified to do the job adequately and of course those will be chosen which are most convenient.

Take the simple case in which the system is just 100 grams of water. Here the state is uniquely defined when only two macroscopic properties have been fixed, and these may be the pressure and temperature. It should be stressed again that the fact that two parameters are needed and not more or less is decided by experience and could not have been predicted *a priori*. In this way it is known that the shape of the specimen of water makes no detectable difference in all ordinary cases; nor does its situation in a magnetic field, or its proximity to a charged conductor. If these things had, however, been of importance it would only have been necessary to specify them to complete the description, and no difference in principle would have been involved.

Now consider the 100 grams in two different states, which will be called (1) and (2). To be definite, suppose that state (1) means that the sample is at laboratory temperature in an open vessel; that is, that it is defined by $T = 20°C$, $P = 1$ atmosphere. Further, let state (2) be realized when the water has been converted into steam in a simple cylinder with a frictionless piston; in this case it is specified as having $T = 110°C$, $P = 1$ atmosphere. (Implicit in these specifications are of course the measures of many other properties, in particular volume, and two of these could have been used alternatively in the description, as mentioned earlier.) The next consideration[1] is how the system can be changed from one state to the other. In the simple case considered it is obviously not a difficult thing to do this; but with more complex cases (such as where the two states concern a living and a dead organism) the transition may be prodigiously hard to carry out. Nevertheless from the thermodynamic point of view it is no more involved, in principle, than the case we are considering. Both concern the same fundamental possibilities; let us consider what these are.

In the first place, since the matter in our system is fixed, all we are concerned with as students of thermodynamics is the addition or removal of different forms of energy. In the second place, we need only consider this energy as transferred in two ways, as heat or as work (we limit ourselves to these because we have seen that all forms of work are interchangeable [2] without loss and so can be obtained from each other). In the third place, it is fairly obvious that there are in fact an infinite number of ways, in general, in which we can communicate the energy, even grouping all the different ways

---

[1] The metaphor of a 'pathway of change' will be frequently used for the temporal sequence of states through which the system passes.

[2] For instance, the mechanical energy of a coiled spring can be converted into electrical energy by means of a dynamo. If work has to be done against electrical forces therefore a coiled spring can be used to do it, in principle without loss.

of doing work together as one. To make this quite clear let us be a little more particular. One way of accomplishing the change would be to place the water, in a suitable container, over a bunsen flame or in contact with a heated metal plate. It would absorb heat, rise in temperature, evaporate and finally undergo heating as vapour until it was at 110°C, the required end point. Another way of proceeding would be to stir the water vigorously. This heats it in a similar way, but we must not lose sight of the fact that we are really doing work on the system, not causing it to absorb heat; the proof being that the energy we put into it is at once calculable (with a little effort) by the ordinary prescription of force times distance. The fact that the work we do on it is at once converted frictionally to 'heat' is, if we may express it so, not our business. What *we* are supplying is work, not heat, and what happens to the energy supplied is for present purposes immaterial. Exactly the same is true if we use an electrical resistance method of heating; what is supplied to the water (or rather to the element) is electrical work (calculable in the usual way, but normally estimated by a derived method using amperes and volts). Finally, of course, we could use partly one method and partly the other, supplying some of the energy as heat from a hotter body and some as work electrically or otherwise. It is here that one possibility of infinite variation occurs, by alteration of the proportions in which we supply energy in the two ways.

When we have carried out the transition from one state to the other we are left with this result: a quantity of heat Q has been supplied to the water, and an amount of work W done upon it. Together these constitute an amount of energy $(Q+W)$ which has been imparted to the system. This is where the First Law becomes applicable, for if energy can neither be created nor destroyed the sum $(Q+W)$ *must be the same whatever the procedure adopted to carry out the transition*. Further if the system is restored from state (2) to state (1) again, the quantity $(Q+W)$ must still be numerically the same, but with the opposite sign. Clearly, if these results were not true the process could be operated cyclically, going from (1) to (2) and back again repeatedly using different paths if necessary, and either manufacturing, or destroying, as much energy as required. That this cannot be done means that the quantity $(Q+W)$ is independent of everything except the initial and final states. This implies in turn that it is not nonsensical to regard the system as possessing in itself a content of energy U, whose change between the two states is given by

$$U_2 - U_1 = \Delta U = Q + W. \tag{2.1}$$

The First Law therefore enables the existence to be established of a function U, called the internal energy, or more simply the energy of the system, whose absolute magnitude is indeed unknown but whose changes are completely defined in the way described. Such a function is called a variable of state, since it is completely fixed when once the state of the system is specified, the previous history of the system being irrelevant. It is therefore a property of the system in just the same sense as say the volume is, the main difference being that unlike volume its absolute magnitude is unknown; but it 'belongs' to the system in precisely the same way.

Before leaving the subject of this section there are two or three points worth emphasizing. The argument employed to establish U makes it clear that there would be no meaning in enquiring about the content in a system of either heat [1] or work. The amount of heat or work added to or abstracted from the system would depend not only on the two states (1) and (2) but also on the path of transition between them, and either could have almost any value we chose. The next point is that in calculating the amount of work W done on the water in this example, account must be taken of the work unavoidably done by the water in expanding and pressing back the atmosphere. This represents an acquisition of energy by the atmosphere (which is lifted a little bit higher), and so must constitute a loss of energy, in the form of work, from the amount supplied to the water. If ΔV is the increase in volume of the water as it changes to vapour, and P is the pressure of the atmosphere, the amount of work done by the water is $P\Delta V$, and the contribution to the increase in energy of the water is the reverse of this, namely $-P\Delta V$. This has to be incorporated, algebraically, in the term W. This matter of including all energy changes in the summation, and with their right signs, is most important. It is easy to understand that when a stretched spring alters in length it has given up, or acquired, a certain increment of potential energy; but it is easy to forget that when a less familiar system, necessarily held at some sort of pressure, changes its volume the same is true; this increment must always be included in the computation.

Two further points are as follows. To make clearer the fact that the number of paths between states (1) and (2) is unlimited the possibility can be mentioned that varying the relative amounts of heat and work supplied in the way previously suggested is not the only one. The system can also be held in many different ways while the energy is being supplied. The water, for instance, might be enclosed in a strong bottle while heating is in progress, the bottle only being opened when enough energy had been imparted. Alternatively it could be vaporized under lower pressure first, and then heated and compressed to give the correct final conditions. The variations on this theme are in fact limitless, and they give rise to an interesting question: is it always necessary to supply (using the term algebraically to cover also 'remove') both heat *and* work if a transition is made from one state to another, or can

---

[1] The function called the 'heat content' or enthalpy (H) has quite a different meaning (see Chapter 8).

this be accomplished with a source of work alone, or with a supply of heat alone? This question goes rather beyond the bounds of the present chapter, but it may be said that the transition can always be accomplished by work alone in at least one [1] of the two directions (i.e. 1→2 or 2→1) and in special cases in both. This means that the possibility always exists of finding experimentally the ΔU concerned by work measurements alone, electrical ones being especially convenient; it is only necessary to select the right direction for the transition. This possibility is very important both practically and theoretically. In the foregoing discussion the quantity of heat Q has been spoken of rather glibly; but in actual fact the direct determination of this quantity may be a very difficult matter, in principle no less than in practice. Where the heat is supplied by say a block of hot metal the measurement is very simple; but where it comes from what physiologists might call a 'metabolizing' system (like a bunsen flame) it is very problematical, even in theory. Thus it is of great importance that the necessity of direct heat measurements can be side-stepped altogether. The notion of heat in fact appears deceptively simple; it has borrowed some of the reputation of temperature as being intuitively elementary.

It cannot be said of heat, as of work, that there is always one direction for a transition in which the other member of the pair can be dispensed with. For instance where states (1) and (2) involve a difference in volume there is always the quantity $\pm P\Delta V$ on the 'work' side, the system either giving or accepting this quantity of work from the surroundings according as we move from (1) to (2), or (2) to (1). On the other hand, when a simple gas is heated at constant volume no work term enters, nor does it when the gas is re-cooled to its original state, so that evidently it cannot be said that work exchanges are invariably necessary. Perhaps this may be summarized as follows: when a system is to be changed from one state to another it will often be the case that both heat and work must be provided (algebraically) to secure the desired result; but for one direction of the change it is always possible to effect the transition using work only. The reason for this state of affairs is connected however with the Second Law, and so lies outside the scope of the present chapter.

Properties of the Internal Energy U

In this section the important properties of the internal energy U will be recapitulated and a few further remarks added.

In accordance with the way in which it has been defined the existence of the function U is a consequence of the First Law, and in so far as its use and

---

[1] In the example discussed the water could be vaporized (1→2) by doing work alone, say in an electric kettle. But for the reverse change (2→1) at least some of the energy must be removed as heat. The reader should think this out.

2. THE FIRST LAW OF THERMODYNAMICS

development leads to consistent and valid results the First Law is confirmed This has so far invariably been the case, even with living systems. The internal energy of a closed system can be increased in only two ways: by doing work on the system (such as by compressing, stretching or distorting it, or by some sort of application of friction to it), and by placing it in contact with a hotter body so that it absorbs heat.

The internal energy is clearly an *extensive* property of the system; that is, its magnitude depends on how large the system is. If two systems are considered together their total internal energy is the sum of their individual values; and as a corollary, the internal energy of an *open* system (i.e. one into which matter can pass from the surroundings, or vice versa) can also be increased by the addition of matter.

Verbal Statements of the First Law

Since it is always a help in grasping an abstract principle to have it stated in alternative ways, two equivalent statements of the First Law are added below:

(1) In an *isolated* system the sum of all forms of energy, including mechanical, chemical and thermal, remains constant.

(2) In any transformation taking place in a *closed* system the increase in internal energy is equal to the work done on the system added to the heat absorbed by it.

What is meant by a closed system has already been explained; an isolated system is one in which in addition to matter, energy also is unable to enter or leave. It is, in fact, completely cut off from all external interaction. In the first statement above three forms of energy important in biological systems are mentioned by name, but there are of course others.

A Caveat

In subsequent chapters equation 2.1 will be used in the differential form (that is, in a form applicable to infinitesimally small changes):

$$dU = dQ + dW. \tag{2.2}$$

The quantities dQ and dW are not the differentials of properties of the system, as dU is. Consequently many writers prefer to denote them by q and w, while others use a special notation like $đQ$ and $đW$. The point is important because while differential coefficients like $\dfrac{dU}{dT}$ stand for definite properties of a material system (in this case its heat capacity) a differential coefficient like

$\dfrac{dQ}{dT}$ has no such significance and serious errors can be made if one thinks it has. In this book the symbols dQ and dW are used, since the quantities are, after all, differentials. What must be remembered is that their magnitude depends on factors in addition to those which suffice to describe the change in the system; in short, on the whole *pathway* of that change. What the First Law implies is that these 'pathway factors', as they may be called, influence dQ and dW in exactly opposite ways, so that when these differentials are added their effect disappears, and the sum (dU) is independent of the pathway.

CHAPTER 3

Some Mathematical Topics

> The word 'lesson' came back to Pooh as one he had heard before somewhere.
> 'There's a thing called Twy-stymes', he said. Christopher Robin tried to teach it to me once, but it didn't.'
> 'What didn't?' said Rabbit.
> 'Didn't what?' said Piglet.
> Pooh shook his head.
> 'I don't know,' he said. 'It just didn't. What are we talking about?'
>
> THE HOUSE AT POOH CORNER
> A. A. Milne

The mathematical equipment necessary to understand thermodynamics is really very slight, although the wealth of symbols, many with both superscripts and subscripts, the very frequent (in fact almost exclusive) use of partial differential coefficients and the functional notation often used, are apt to be very frightening to the biologist who has had little time to spare for acquiring mathematical skill. This chapter is therefore added to make this side of the subject less of a stumbling block.

FUNCTIONAL NOTATION

Firstly functional notation is introduced. The expression

$$y = f(x) \tag{3.1}$$

is probably well understood already. It is simply a shorthand way of saying that the variable quantity y depends on the variable quantity x in the sense that if x is given a particular value the value of y is thereupon fixed. Of course the reverse would also be true; giving a value to y would mean that the value of x had thereby been settled. The relationship between x and y can be conveniently represented by a two-dimensional graph (Fig. 3.1) in which values of x are measured parallel to OX as distances from OY, and values of y parallel to OY as distances from OX, the two axes OX, OY being almost invariably perpendicular to one another. The reciprocal relationship between x and y means graphically that if a value OM of x is fixed, the corresponding value ON of y is implied by the graph QPR of the function in the way shown; the same applies when the value ON of y is settled first. It is a very general convention that the variable which is fixed first (the 'independent variable') has its axis horizontal, the other (the 'dependent variable') being plotted vertically; but when necessary this convention can be reversed, and the variables even regarded each in the other role (that is, y as the independent variable and x as the dependent one). The symbol f may stand for some well-known mathematical function like 'the square of', or 'the log of', or 'three

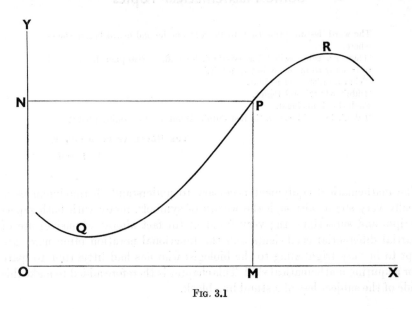

Fig. 3.1

times the exponent of, plus seven times the first power of, plus six', these particular examples being written symbolically

$$y = x^2$$
$$y = \log x \qquad (3.2)$$
$$y = 3e^x + 7x + 6.$$

However, it has to be remembered that the statement that y is a function of x, that is $y = f(x)$, does not imply that the mathematical formula connecting them can be written down. Very often when dealing with thermodynamic quantities it is found that the relationship between them is far too complex to make it fit any simple mathematical expression. For an ideal gas at constant temperature

$$P = c/V \qquad (3.3)$$

and the function $f(V)$ is the simple one 'c times the reciprocal of V'; but for other classes of substance the relationship cannot be expressed exactly. In other words, the form of f is not known, but in principle that makes no difference at all. The general formula (3.1) is still applicable and the relationship can still be represented by a two-dimensional graph.

It sometimes happens that we do not wish to imply that either of the variables x and y is any more 'independent' than the other, and in this case instead of using the relation (3.1), it may be written in the symmetrical form

$$F(x, y) = 0 \tag{3.4}$$

which is to be interpreted in precisely the same way. The symbol F, incidentally, has been used here instead of the previous f merely to avoid the suggestion that the function has exactly the same form.

Finally, where the value of a variable z, depends on the simultaneous values of two others, x and y, we write

$$z = f(x, y) \tag{3.5}$$

or

$$F(x, y, z) = 0. \tag{3.6}$$

As an example of this the equation of state for a simple substance can be quoted as

$$V = f(T, P) \tag{3.7}$$

or

$$F(V, T, P) = 0 \tag{3.8}$$

where V, T, P have their usual meanings. A usage which is often met with in thermodynamics (where so many alternative independent variables can be chosen) is to employ symbols like $V(T, P)$. This stands for the volume; but it carries the additional implication that the volume is here being considered as dependent on the temperature and pressure.

DIFFERENTIAL COEFFICIENTS

There are two rather obvious features of a graph such as Fig. 3.2 which there will often be cause to use. One strikes the eye immediately; it is the slope or gradient. This is represented graphically by the tangent of the angle θ in the figure. A knowledge of its value answers the question: as x increases, how fast relatively does y increase? When the gradient is large, y is clearly very sensitive to changes in x; and when it is small, it is insensitive. One particular case is of great importance. When the curve reaches a maximum or minimum (R or Q, Fig. 3.2) the gradient becomes zero, and this means that for the moment, as x passes through these values, y becomes completely insensitive. It is said to be 'stationary'. Such points, as will be seen later, have a particular relevance to equilibrium. The gradient, of course, except in the special case where the graph is a straight line, changes from point to point; and to meet this situation the mathematician measures it by the 'differential coefficient' $\dfrac{dy}{dx}$, which will now be explained. If x increases by an amount δx (which may be quite large), and y increases concomitantly by the amount δy, then the quotient $\dfrac{\delta y}{\delta x}$ clearly measures the *average* gradient during the change (Fig. 3.3). To call this quotient the gradient *at* P would obviously be to speak very loosely in more ways than one; for instance,

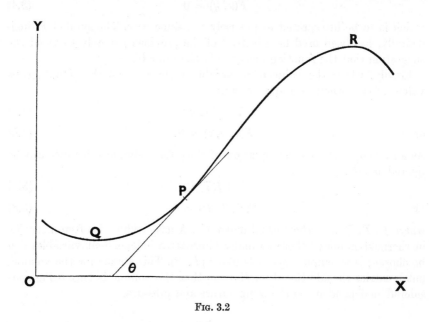

Fig. 3.2

the value of the quotient would have no preciseness about it, since it would depend on how big we took the increment δx. If, however, δx (and so δy) become smaller and smaller it is found that their ratio $\dfrac{\delta y}{\delta x}$ in all ordinary cases tends to get nearer and nearer to a quite definite value, and this value is called the gradient at P precisely. Under these circumstances $\dfrac{\delta y}{\delta x}$ is written $\dfrac{dy}{dx}$, and for non-rigorous purposes this expression may be regarded as a fraction whose numerator and denominator are exceedingly small, corresponding increments of y and x called differentials. It may be treated (though the mathematician would frown at this) just as any other fraction, cancelling the quantities dx and dy where appropriate, and separating them where necessary. Regarding them in this way as quite ordinary, though extremely small, quantities will not be harmful; and it will lead at once to the recognition of many relations which otherwise might present problems. Thus regarding the differential coefficient (let us call it for the moment p) as an ordinary quotient gives

$$p = \frac{dy}{dx}$$

or
$$dy = p \cdot dx \tag{3.9}$$

which explains the name 'differential coefficient', it being revealed as the coefficient by which the differential dx has to be multiplied in order to give the corresponding differential dy.

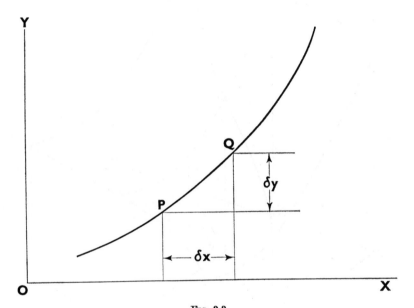

FIG. 3.3

Further it can be seen at once that $\frac{dx}{dy}$ is the reciprocal of $\frac{dy}{dx}$; and if z is a function of y and y is a function of x, then

$$\frac{dz}{dx} = \frac{dz}{dy} \cdot \frac{dy}{dx} \tag{3.10}$$

by the ordinary rule for multiplying fractions. At the extreme top or bottom of a curve (R or Q, Fig. 3.2) where the gradient is zero

$$dy = 0 \cdot dx$$
$$= 0. \tag{3.11}$$

Partial Derivatives

It is very frequently the case that a variable quantity (say z) depends not merely on one other variable, but on two (say x and y). This can be expressed as before by writing

$$z = f(x, y) \tag{3.12}$$

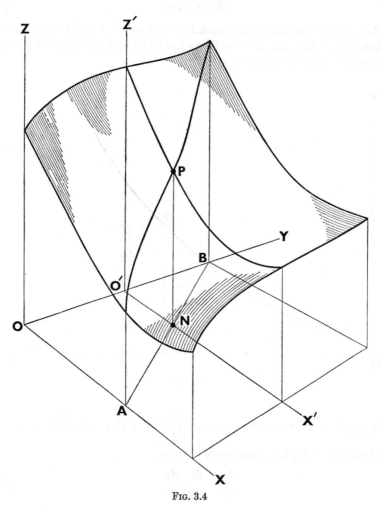

Fig. 3.4

and the relationship can be represented graphically by means of a solid model. The two independent variables x, y may be set out as ordinary rectangular co-ordinates on a baseboard, the dependent variable z being measured vertically upwards above the board. The result will be that the relationship expressed by equation (3.12) is represented by a surface, a sort of undulating 'hillside' (Fig. 3.4). If it is necessary to describe how the 'land lies' at a particular point P, it is found that now not one but two gradients have to be specified (or at least two equivalent quantities). The most usual and convenient way of doing this is to imagine two vertical planes to pass through P, one perpendicular to the direction in which y is measured, and the other perpendicular to the direction in which x is measured. These will intersect the 'hillside' in two curved lines each passing through P. All along the first one y will naturally have a constant value, and all along the second the same will be true of x. The first of these planes is shown in Fig. 3.4 as X'O'Z', and in Fig. 3.5 it is shown with its line of intersection with the hillside

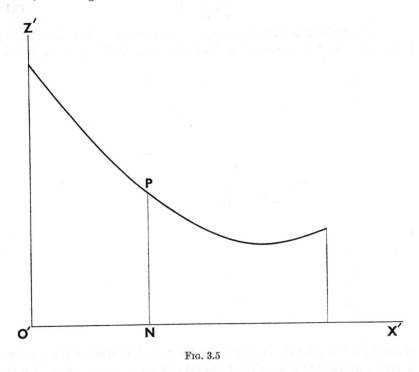

Fig. 3.5

isolated by itself. The slope of this line at P is one of the quantities needed to represent the 'lie' of the surface at P; and it is written not $\dfrac{dz}{dx}$, but $\left(\dfrac{\partial z}{\partial x}\right)_y$. The reason for this slight change of notation is to emphasize firstly that z does not depend on x only (by using a special type for the letter d); and secondly to indicate what is happening to the other variables on which z depends while x is varied. In this particular case there is only one other independent variable, and that is kept constant when x changes; this is indicated by the small subscript y. In an analogous way the 'partial derivative' $\left(\dfrac{\partial z}{\partial y}\right)_x$ may be formed; this is the slope of the line of intersection with the surface of a vertical plane perpendicular to the x-axis (this plane has not been drawn in). It is important to note, however, that we are not limited to these two gradients. As an example of this, suppose there is a vertical plane through P cutting the surface in the line shown, and cutting the axes OX, OY at A, B (where $x = a$, $y = b$, Fig. 3.4). All over this plane it is clearly the case that

$$\frac{x}{a} + \frac{y}{b} = 1 \qquad (3.13)$$

(the reader unfamiliar with co-ordinate geometry can easily verify that this is the equation of the straight line AB in the plane XOY, and so in the solid

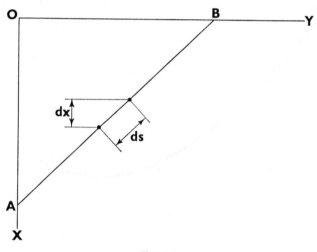

Fig. 3.6

of the plane through this line parallel to the z-axis). Further, if this equation is varied by replacing the right hand side by an arbitrary quantity u the whole family of vertical planes parallel to this one is obtained. Thus the partial differential coefficient $\left(\dfrac{\partial z}{\partial x}\right)_u$ will be related to the slope of the 'hillside' when cut by one of these vertical planes oriented to both the x- and the y-axes. It will not be the slope exactly, because its denominator (dx) is not measured *along* this plane. In order to correct for this ds should be substituted for dx (see Fig. 3.6, which represents the basal plane XOY). The point is not important in the present connection, however; what is important is that the partial derivatives $\left(\dfrac{\partial z}{\partial x}\right)$ and $\left(\dfrac{\partial z}{\partial y}\right)$ can be formed not only holding y and x constant, but maintaining constant any function of them both (say $u(x, y)$ that may present itself.

In general terms this can be amplified as follows. The equation[1]

$$u(x, y) = \text{constant}$$

(where u by definition does not contain z) represents a vertical surface standing on the xy plane and intersecting it in the curve

$$u = \text{constant}.$$

This vertical surface cuts the hillside in a line.

We take two points P_1, P_2 on this line an infinitesimal distance apart, and consider the changes dx, dy, dz in their co-ordinates. The quotients $\dfrac{dz}{dx}$ and $\dfrac{dz}{dy}$ are formed; these are of course the partial derivatives $\left(\dfrac{\partial z}{\partial x}\right)_u$ and $\left(\dfrac{\partial z}{\partial y}\right)_u$. It is easy to see that the partial derivatives in which y and x are held constant are merely particular cases of these; in one u is simply equal to y, and in the other to x. Figure 3.7 illustrates these relationships.

Total Differential of z

This leads to a rather important point. Suppose z is a function of x and y in the way discussed; that is

$$z = f(x, y). \tag{3.14}$$

Then if x increases by the infinitesimal (for our purpose excessively minute, but finite) amount dx, and y correspondingly increases by the infinitesimal amount dy, by how much does z increase? In the case of $y = f(x)$ the corresponding answer was

$$dy = p \cdot dx \tag{3.15}$$

and this relationship gives the name 'differential coefficient' to p. In the present case the corresponding result is

$$dz = \left(\dfrac{\partial z}{\partial x}\right)_y \cdot dx + \left(\dfrac{\partial z}{\partial y}\right)_x \cdot dy \tag{3.16}$$

---

[1] Thus if $u \equiv x^2+y^2$, the curve $u = c$ is a horizontal circle centre O and radius $\sqrt{c}$, and the surface $u = c$ is a right cylinder standing on this. The symbol $\equiv$ means that u is here *another name* for (x^2+y^2); that is, identically equal to it, and not just equal to it for some particular values of x and y.

a result which can be fairly readily appreciated intuitively. Geometrically, it means that the total distance risen in moving on the hillside can be found by making the transition in two stages. This is done by moving from P_1 to Q

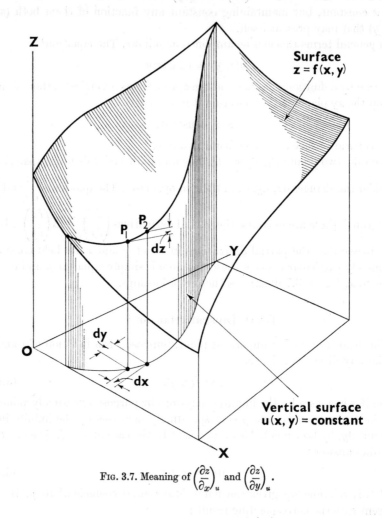

Fig. 3.7. Meaning of $\left(\dfrac{\partial z}{\partial x}\right)_u$ and $\left(\dfrac{\partial z}{\partial y}\right)_u$.

(Fig. 3.8) at constant y and rising by the amount $\left(\dfrac{\partial z}{\partial x}\right)_y . dx$; then moving at right angles from Q to P_2 at constant x and rising a further amount $\left(\dfrac{\partial z}{\partial y}\right)_x . dy$. Each of these increments is only part of the total increment in z; hence the derivatives are spoken of as 'partial differential coefficients', in line with what was mentioned earlier. In equation (3.16) it must of course be remembered that the differentials of the independent variables must each be multiplied by the right partial derivative; and the partial derivatives must be held constant each time with respect to whatever variable is going to be introduced in a later term as a differential itself. This is merely to say that in passing from P_1 to Q and then to P_2 (Fig. 3.8) the horizontal distance

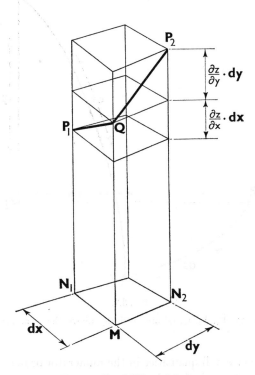

FIG: 3.8. Total differential of z.

N_1M must be multiplied by the slope of P_1Q, and the horizontal distance MN_2 by the slope of QP_2, where N_1M and P_1Q are in the same vertical plane, and likewise MN_2 and QP_2. Provided this condition is met it does not in fact matter how the distance from P_1 to P_2 is moved. In the plan (Fig. 3.9) the movement can be parallel to the axes of x and y in the two steps dx and dy; or alternatively along other systems of lines P_1Q', $Q'P_2$ in steps ds and dt. If this is done then the partial derivatives $\left(\dfrac{\partial z}{\partial s}\right)_t$ and $\left(\dfrac{\partial z}{\partial t}\right)_s$ will have to be used to multiply them.

Manipulation of Partial Derivatives

In just the same way as in the discussion of ordinary differential coefficients partial derivatives can be regarded (if we keep quiet about it) as simple

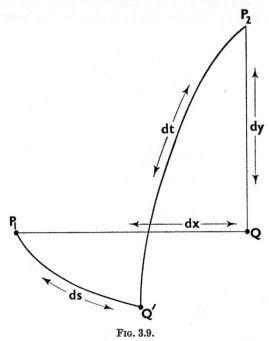

Fig. 3.9.

Note that the points P_1, P_2, Q, Q' are vertically above the positions shown (compare Fig. 3.8).

fractions with very small quantities in the numerator or denominator. Thus they can be turned upside down[1]:

$$1 \bigg/ \left(\frac{\partial z}{\partial x}\right)_y = \left(\frac{\partial x}{\partial z}\right)_y \qquad (3.17)$$

the subscript obviously remaining unchanged. They can also be multiplied together, cancelling numerator with denominator in the usual way, but only when the subscripts are the same:

$$\left(\frac{\partial z}{\partial x}\right)_y \cdot \left(\frac{\partial x}{\partial t}\right)_y = \left(\frac{\partial z}{\partial t}\right)_y. \qquad (3.18)$$

[1] This cannot be done (for reasons which will appear on reflection) with the higher derivatives $\frac{d^2y}{dx^2}$, $\frac{\partial^2 z}{\partial x^2}$, etc.

But a moment's reflection will show that there is nothing immediate that can be done with such a product as

$$\left(\frac{\partial z}{\partial x}\right)_y \cdot \left(\frac{\partial x}{\partial t}\right)_s$$

where the subscripts are different. However, an interesting result can be obtained directly from the fundamental equation (3.16). Imagine a horizontal plane to pass through the point P (Fig. 3.4), cutting the surface in contour lines, for which z is constant. Considering two points very close together on one of these, then

$$dz = 0$$

the relationship between dx and dy for the infinitesimal separation being obtained by writing $dz = 0$ in equation (3.16). This gives

$$\left(\frac{\partial z}{\partial x}\right)_y \cdot dx + \left(\frac{\partial z}{\partial y}\right)_x \cdot dy = 0 \qquad (3.19)$$

i.e.

$$\frac{dy}{dx} = -\left(\frac{\partial z}{\partial x}\right)_y \bigg/ \left(\frac{\partial z}{\partial y}\right)_x. \qquad (3.20)$$

This result presupposes that the left hand side is measured under conditions in which z is held constant. Incorporating this requirement explicitly, $\frac{dy}{dx}$ is accordingly replaced by $\left(\frac{\partial y}{\partial x}\right)_z$; and rearrangement of equation (3.20) into a symmetrical form yields the important and easily remembered result

$$\left(\frac{\partial z}{\partial y}\right)_x \cdot \left(\frac{\partial x}{\partial z}\right)_y \cdot \left(\frac{\partial y}{\partial x}\right)_z = -1 \qquad (3.21)$$

in which it appears, deceptively, permissible to cancel numerators and denominators in spite of the differing subscripts. That it is deceptive is obvious since this procedure would not yield the negative sign. It is an interesting geometrical exercise to prove this equation from first principles.

Conditions for a Maximum or Minimum

When dealing with maxima and minima ('stationary values') in the simple case of $y = f(x)$ it was seen that the criterion for these is that $dy = 0$ for infinitesimal values of dx. In other words, the differential coefficient $p = \frac{dy}{dx}$

must be zero. In the more usual thermodynamic case, where there are at least two independent variables, the condition is similarly dz = 0 for any very small arbitrary changes dx and dy (dz is of course also zero along a contour line, but dx and dy cannot then both be arbitrary). Since dx and dy are independent (e.g. one can be held at zero while the other is given a test value) and since equation (3.16) must hold, it follows that for an absolute maximum or minimum both partial derivatives must be simultaneously zero. If only one partial derivative is zero the point being considered will be merely on a 'crest'; it needs *both* to be zero simultaneously before the 'summit' is reached. (These statements apply of course to a maximum; 'valley' and 'bottom' would have to be substituted for a minimum.) Correspondingly, when only one partial derivative is zero a *partial* maximum or minimum is spoken of; when both are zero the stationary values are said to be *absolute*. This terminology will later be applied to equilibria.

More than Two Independent Variables

Consideration need not, of course, be limited to the case of two independent variables; often in this subject considerably more are involved. Thus it may be that

$$w = f(x, y, z) \tag{3.22}$$

and there will then be three partial derivatives in each of which only one variable will be considered to vary. They are written

$$\left(\frac{\partial w}{\partial x}\right)_{y,z}, \quad \left(\frac{\partial w}{\partial y}\right)_{z,x}, \quad \left(\frac{\partial w}{\partial z}\right)_{x,y}$$

and the equation for the total differential of w is

$$dw = \frac{\partial w}{\partial x} \cdot dx + \frac{\partial w}{\partial y} \cdot dy + \frac{\partial w}{\partial z} \cdot dz \tag{3.23}$$

where the subscripts have been omitted for brevity. The geometrical interpretation becomes in such cases rather more difficult, but no new principles are involved.

Differential Coefficients of Higher Orders

The further extension of the idea of differential coefficients to higher orders is probably already familiar to the reader. Referring to an earlier figure (Fig. 3.2) the gradient at P is itself a function of x. Another graph can be plotted in which, not y, but this gradient $p = \dfrac{dy}{dx}$ is set up in the vertical direction (Fig. 3.10). This curve is spoken of as the first derived curve of the previous one, and it will of course have a gradient $\frac{dp}{dx}$ of its own. This gradient is given the rather artificial notation $\frac{d^2y}{dx^2}$, and the process can obviously be repeated to give $\frac{d^3y}{dx^3}$ and still higher derivatives. When considering a function of two independent variables the possibilities are even more numerous. The derivative $\frac{\partial z}{\partial x}$ can be 'differentiated' again with respect to x to give $\frac{\partial^2 z}{\partial x^2}$,

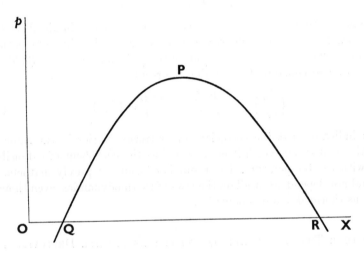

Fig. 3.10

or it can be differentiated with respect to y to give $\frac{\partial^2 z}{\partial x \partial y}$, keeping the appropriate variables constant in each case. A very useful theorem makes its appearance here, called the cross-differentiation identity. It states that in whichever order the two partial differential coefficients are formed the result will be the same. In symbols,

$$\frac{\partial^2 z}{\partial x \partial y} = \frac{\partial^2 z}{\partial y \partial x}. \tag{3.24}$$

This is quite generally true for all functions except for what might be called mathematical freaks, which are not considered here. It can be interpreted geometrically, though the interpretation is not particularly helpful; but it can lead to some very useful results. Consider for instance the properties of a steel wire. The coefficient of expansion (α) of the metal can be defined as

$$\alpha = \frac{1}{L}\left(\frac{\partial L}{\partial T}\right)_F = \left(\frac{\partial \ln L}{\partial T}\right)_F \tag{3.25}$$

where the length L[1] is measured at a variable temperature T, but naturally under constant tension F. On the other hand, the mechanical property of extensibility (ϵ), which measures the proportional change in length under load, is expressible by the relation

$$\epsilon = \frac{1}{L}\left(\frac{\partial L}{\partial F}\right)_T = \left(\frac{\partial \ln L}{\partial F}\right)_T \tag{3.26}$$

where in a similar way temperature is presumed to be maintained constant. The cross-differentiation identity can now be applied by differentiating equation (3.25) with respect to F at constant T, and equation (3.26) with respect to T at constant F. Equating the results gives

$$\left(\frac{\partial \alpha}{\partial F}\right)_T = \frac{\partial^2 \ln L}{\partial T \partial F} = \frac{\partial^2 \ln L}{\partial F \partial T} = \left(\frac{\partial \epsilon}{\partial T}\right)_F \tag{3.27}$$

which indicates that the coefficient of expansion varies in the same way with the load under which it is measured as the coefficient of extensibility does with the temperature. The result is of course a purely mathematical one and not dependent at all on the laws of thermodynamics, even though it concerns thermodynamic quantities.

Other Thermodynamics Quantities as Partial Derivatives

Partial derivatives occur in thermodynamics far more often than ordinary ones for the simple reason that a thermodynamic system is characterized by so many different variables. Even in the simplest of cases, such as that of a perfect gas, there are three (usually P, V and T). It is therefore important for the reader to be able to express ordinary properties of systems as partial derivatives, or alternatively to be able to interpret in terms of simple operations what is the meaning of a partial derivative he has not met with before.

Consider, for instance, the meaning of such a derivative as $\left(\frac{\partial U}{\partial T}\right)_V$. This measures how the internal energy of the system changes with temperature when the volume is maintained constant. For an ordinary system (i.e. when such special forces as magnetic or electrical ones are excluded) constancy of volume implies that no work is done on or by the system, hence its energy

---

[1] The symbol $\ln L$ stands for the natural logarithm of L, i.e. $\log_e L$.

can only be increased by the absorption of heat. The derivative therefore measures how much heat has to flow into the system to raise its temperature by a given amount. In other words, it is an ordinary heat capacity, which, if divided by the mass of the system, would give specific heat. It is assumed of course in all this that the system is a closed one, and if it contained n moles of matter this proviso could be made quite explicit by adding n as an additional subscript, $\left(\dfrac{\partial U}{\partial T}\right)_{V,n}$, though this is usually quite unnecessary.

The reminder that all systems are not closed introduces a very important class of partial derivatives known as partial molar quantities. These become important whenever systems are considered that can exchange matter as well as energy, so-called open systems. Consider, for instance, a simple solution of two components, solvent and solute. If a volume v of solute be added to the solution it can readily be appreciated that the volume of the latter will not be increased by exactly v. Put into other words the overall volume of dry solute and solution may undergo a change, positive or negative, when the two mix. Dry gelatine and water are a well known case where there is a marked overall shrinkage. Since volume is therefore not conserved while the mass is, it becomes important to express the way in which the volume of a system increases when a specified sort of matter is added to it. Suppose the volume of the solution being considered is V, and that it contains n_1 moles of solute and n_2 moles of solvent. Then the partial derivative $\left(\dfrac{\partial V}{\partial n_1}\right)_{n_2}$ obviously expresses the required measure. The subscript n_2 expresses the fact that in thought an operation is being carried out on the system without altering the amount of solvent present. Adding dn_1 moles of solute, the amount dV by which the volume of the solution increases is noted and then dV is divided by dn_1 to obtain the amount by which the solution would increase in volume per mole of added solute if it did not alter qualitatively in the process, that is, if it was large enough not to feel the addition. Naturally the temperature and pressure must be maintained at a constant level throughout or the result would have no unique value. Hence the quantity

$$\bar{V}_1 = \left(\dfrac{\partial V}{\partial n_1}\right)_{T,P,n_2}$$ known as the partial molar volume of the solute (1) is obtained. The *molar* volume of the solute is merely the volume per mole of the pure substance, and in ordinary cases the *partial* molar volume would not be far different from it.

The idea of partial molar quantities can clearly be extended to cases where there are more than two components present (suitable extra subscripts n_3, n_4, etc., are merely added to the derivative) and also to extensive quantities other than volume. A particularly important case is that of the Gibbs free energy G, where the partial molar quantity $\left(\dfrac{\partial G}{\partial n_1}\right)_{T,P,n_2}$ has highly important properties. It is given the special name of the chemical potential of the component in question, and a symbol of its own (μ). It will be seen later that under conditions of constant temperature and pressure there is a natural urge for the Gibbs free energy of the system to decrease. The partial molar quantity $\dfrac{\partial G}{\partial n_1} = \mu_1$ measures how rapidly the free energy falls when component (1) is withdrawn; consequently it is not difficult to appreciate that μ_1 measures the tendency of this component to escape. It is because of this that this partial molar quantity has such great importance, especially in chemical reactions and membrane equilibria.

Before leaving the subject of partial molar quantities a relation must be mentioned which will later prove of value. It will be stated for the Gibbs free energy (G) and for the volume, but it is also valid for any other extensive property (e.g. entropy or internal energy), and for a system of three components though it is true for any number. The relation in question is

$$G = n_1 \frac{\partial G}{\partial n_1} + n_2 \frac{\partial G}{\partial n_2} + n_3 \frac{\partial G}{\partial n_3}$$
$$= n_1 \mu_1 + n_2 \mu_2 + n_3 \mu_3 \tag{3.28}$$

for the Gibbs free energy, and

$$V = n_1 \frac{\partial V}{\partial n_1} + n_2 \frac{\partial V}{\partial n_2} + n_3 \frac{\partial V}{\partial n_3}$$
$$= n_1 \bar{V}_1 + n_2 \bar{V}_2 + n_3 \bar{V}_3 \tag{3.29}$$

for the volume. It should be emphasized that these equations are purely mathematical ones and do not depend on the laws of thermodynamics at all. A simple way of visualizing their truth can be given as follows using the volume as illustration.

Consider the volume V of a three component mixture. Imagine a very small quantity dn_1 of the first component withdrawn. The reduction in volume is naturally $\bar{V}_1 dn_1$, by definition of $\bar{V}_1$. If more of this component is withdrawn then $\bar{V}_1$ begins to alter, so instead dn_2 of the second component is withdrawn, producing a further reduction in volume of $\bar{V}_2 dn_2$. The next operation is to withdraw dn_3 of the third component. The result of the first round is therefore to reduce the total volume by an amount given by

$$d\bar{V} = \bar{V}_1 dn_1 + \bar{V}_2 dn_2 + \bar{V}_3 dn_3. \tag{3.30}$$

Now if dn_1, dn_2 and dn_3 have been chosen in the right proportions the mixture will obviously have the same composition as it had to begin with, only there will be less of it. We can clearly repeat the whole round of subtractions over and over again, the volume getting less and less but its composition never changing. Finally it will all have disappeared, each of the components becoming exhausted at the same moment because of the proportions in which they were removed. Thus the whole volume V can be found by summing up a number of equations like (3.30), the important point being that $\overline{V}_1$, $\overline{V}_2$, and $\overline{V}_3$ remain constant. The result is of course relation (3.29).

The Sign of Integration

For the sake of the reader unfamiliar with mathematics, this chapter will be concluded with a brief mention of integration. It was seen earlier that a very obvious property of the graph representing the functional relation

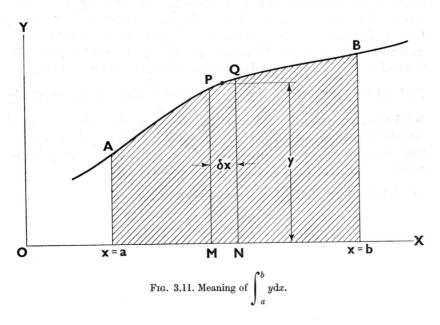

Fig. 3.11. Meaning of $\int_a^b y\,dx$.

between two variables x and y is its slope, and that this may be defined as the value of the quotient $\dfrac{\delta y}{\delta x}$ when the corresponding increments δx and δy become infinitesimally small. Another very obvious property of the curve is the area under it (Fig. 3.11). In thought, this can be measured in the following way. Imagine a vertical strip MNQP marked out as shown, of width δx. If y is the height of the curve at a suitable point between M and N then the area of the strip will be approximately $y\delta x$. If the region AB in which we are interested is divided up in this way then summation[1] of the areas of the strips gives the result

$$\text{Area (A to B)} = \sum_{A}^{B} y\delta x. \tag{3.31}$$

This will, of course, only be an approximation; but just as a precise definition of slope was arrived at by imagining δx to become infinitesimally small, so a precise definition of area may be obtained by the same stratagem. The result is then written

$$\text{Area (A to B)} = \int_{a}^{b} y\,dx \tag{3.32}$$

where the integral sign is an elongated S. In a corresponding way, it is safe in the practical world of biological thermodynamics if the dx is regarded here as being a very small, but finite quantity, and the whole integral as being the sum of an extremely large number of quite ordinary products of the form $y \times dx$. Of course the integral is meaningless if each dx has not got a definite value of y to be multiplied by; in other words if y is not a function of x. But if y is related functionally to x then the symbol $\int_{a}^{b} y\,dx$ stands for a perfectly definite quantity, even if the form of the functional relation is quite out of reach of the mathematician. This is a point which it will be useful to remember when the existence of the thermodynamic property called entropy is discussed.

[1] This is what the symbol $\sum$ (the Greek capital s) means.

CHAPTER 4

Reversibility and Irreversibility

> When Eeyore saw the pot, he became quite excited. He picked the balloon up with his teeth, and placed it carefully in the pot; picked it out and put it on the ground; and then picked it up again and put it carefully back.
> 'So it does!', said Pooh. 'It goes in!'
> 'So it does!', said Piglet. 'And it comes out!'
> 'Doesn't it?' said Eeyore. 'It goes in and out like anything.'
>
> WINNIE-THE-POOH
> *A. A. Milne*

In these days when the high cost of labour has made many people undertake the less skilled type of repair and maintenance job themselves, the process of recharging an accumulator battery will be familiar enough. Some of those who regularly carry out the servicing of their own cars in this respect will even be well acquainted with the performance curves (Fig. 4.1) supplied by

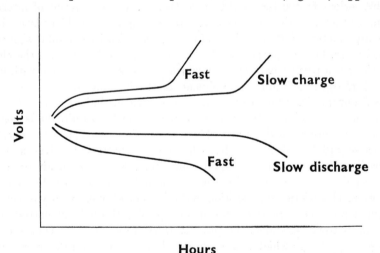

FIG. 4.1. Characteristic curves for battery performance.

the battery manufacturers, curves which show how the battery voltage varies during both the charging and the discharging operations. A universal feature of these curves is that the voltage on charge is always indicated as lying above that on discharge; in other words a higher voltage always has to be applied to the battery to charge it than is obtained from the battery when it is doing useful work. Further, the more rapidly the two operations are carried out the bigger the discrepancy between them. It is fairly obvious that this circumstance means that a loss of useful energy is sustained each time the battery is taken through a cycle; and it is hardly too much to say that it can be appreciated intuitively that this useful energy has been lost irrevocably. Later it will appear that this irrevocability can be established by detailed argument. Charging and discharging an accumulator battery is of course an example of a very widespread phenomenon, the interchange of one form of energy into another (here, electrical into chemical). This process can take very many different forms; for the moment, however, the observation may be summarized by saying that in any practical cycle of operations in which energy is transformed into another form, and then back again, the efficiency falls short of one hundred per cent, and the deficit represents a loss of something useful which is final and irrevocable in the context of the whole universe.

Considering this matter further, why is it that the efficiency can never reach one hundred per cent? The first point that needs to be made clear is that this falling short is not *necessarily* due to imperfection in the design and construction of the battery. As car owners know only too well, some batteries exhibit serious faults; they fail to hold their charge as a result of internal shorts, or they lose their capacity due to disintegration of their plates. But these are practical matters only, and can be remedied by careful design and manufacture. For the purpose of this discussion it can be assumed therefore that there is a perfect battery—one which shows no deterioration with use, holds its charge indefinitely, and delivers exactly the same amount of electricity as is put into it. Why does such a battery fall short in its efficiency?

The answer to this question touches on points which are quite fundamental to the study of thermodynamics. There are in fact two points to consider and these will be taken in order. Charging the battery involves moving electricity through solid and liquid conductors. Now motion always calls into play the forces generally known as frictional, and the present case is no exception. This frictional opposition to the movement of electricity is usually known as *resistance;* and if I is the current passing through the battery and R is the battery resistance the product (IR) of the two, by Ohm's Law, represents a voltage which is used up merely in overcoming this resistance. It has therefore to be *added* to the voltage necessary to effect the charging operation. It is a universal characteristic of frictional forces that they always oppose motion; consequently when the battery is subsequently called on to do useful work the voltage IR has to be *subtracted* from the total available. Frictional effects are on this account spoken of as irreversible. There is a slight verbal confusion here in saying that the frictional forces do in fact reverse or change their direction. Thermodynamic reversibility or irreversibility however does not refer to direction in space, but in time; and it deals not with vector quantities like forces and displacements, but with the outcome of processes. Thus the anomaly occurs that a force which changes its direction when the motion is reversed is associated with thermodynamic irreversibility; while, conversely, forces like weight which do not change direction when the body is moving up or down have thermodynamically reversible effects! However, the paradox is more apparent than real; thermodynamics is interested in the work done by forces, and if both the force and the movement are reversed their product (by ordinary algebra) remains of the same sign.

The forces associated with the electrical resistance of the battery can be made very small by careful design. The plates, for instance, can be placed closer together, and the area they present to their opposite numbers can be increased. It is important to realize that this improvement can take place theoretically without limit; but there are problems of economics, geometry and strength of materials which in fact set a limit to what may be accomplished. However, in principle this source of loss of efficiency can be cut down to any chosen degree, just as the ideal of a frictionless bearing can be approached without meeting any limit set by the constitution of nature. Thus thermodynamic arguments which invoke frictionless mechanisms or cells without resistance are not so out of touch with the world of real things as they may seem; they are merely exploiting the fact that in principle such idealities can be very closely approached, and that in actual engineering practice we often do get quite close to them.

There is, if anything, a deeper reason still why the battery falls short of perfection. Most students are aware of Le Chatelier's Principle, relating to the changes manifested by a system in equilibrium when its constraints are altered. This principle can be extended to systems in which a steady process is in operation; that is, to systems not in equilibrium. It then implies, broadly speaking, that the system alters so as to oppose the process taking place. Now apply this to the charging of the battery. The electric current will of course produce chemical changes at the positive and negative plates. These will result in the electrolyte losing its homogeneity; near the positive plate its composition will deviate in one way from the original and near the negative plate in another way. Thus concentration gradients will be set up between the two plates, and according to the extended principle, these gradients will be such as to *oppose the current*, that is, they will be reflected in a back e.m.f. which will have to be overcome if charging is to proceed. Exactly the opposite state of affairs will occur when the battery is being drawn on. This unwanted e.m.f. thus changes its sign when the current reverses and so, like frictional forces, it is associated with a quantity of work which is always lost, and can never under any circumstances be returned. This constitutes the second reason why the charge and discharge curves do not coincide.

There is however a fundamental difference between the two irreversible effects we have discussed, the frictional effect and that called into play by the extended Principle of Le Chatelier. The makers' curves (see Fig. 4.1) often show the influence of charging and discharging at different rates and when they do so it is always apparent that the faster the operations are carried out the greater is the loss of efficiency in the overall process; in other words the higher the charging curve lies above the discharging one. Frictional forces in a fluid medium (the battery electrolyte) differ from those between solids in that they vary from zero when the velocity is nil to progressively higher values as the velocity increases, whereas between solids they are greater at rest than when motion has been established. Thus a faster battery cycle would mean higher frictional forces and a reduced efficiency, contributing to the effect that has been described. But this is not the whole story, or even the principal part of it. The fact is that the back e.m.f. called into play by Le Chatelier's extended Principle would spontaneously disappear if it is only given the chance to do so. Concentration differences have a natural tendency to even-out; they will disappear of their own accord at no cost in work to ourselves, but they require time to do this. This is where the factor of speed comes in; in a rapid charge or discharge we build up the concentration gradients to a high value, and have to take the consequences in the form of a reduced efficiency and irrevocable loss of useful work. If nothing more were done than conducting the operations more slowly the natural processes of diffusion (which cost nothing) would be allowed to maintain the electrolyte in a more homogeneous state, and as a consequence the irrecoverable loss of useful work would be less.

Such considerations as those mentioned above lead to the conclusion that in principle there is really no reason at all why the two characteristic battery curves should not be entirely coincident, since the frictional element can be reduced by careful design (without theoretical limit); and what might be called the Le Chatelier Effect can be overcome by operating more slowly, again involving no lower limit. When the process of charging or discharging the battery is carried out so slowly that the same voltage curve suffices whichever is the direction of the current it is said to be carried out in a thermodynamically reversible manner. The term 'reversible' is thus given a rather special and precise meaning, which should now be fairly clear. It is not enough that the system—in the present case a battery—should be capable of being restored to its original chemical state. In that sense any change whatsoever is probably reversible, if only we are prepared to pay for it, in terms of energy. What is important is that the change should be carried out in such a way that every step can be *exactly* retraced, all the parameters such as pressure, volume, temperature, voltage, chemical composition and so on being exactly as they were before at the appropriate point on their journey, but changing with time in the opposite direction. For this to be the case it is evident that we must conduct the changes in our system in such a way that it is never for a moment out of equilibrium within itself, or with its surroundings. To use the example of the battery again, suppose lack of internal equilibrium is allowed to occur in the battery under charge. Suppose the point M has been reached on the charging curve (Fig. 4.2) and imagine

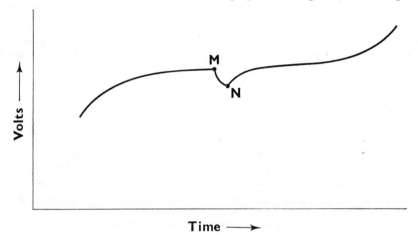

FIG. 4.2. Interrupted charging of a battery.

the charger to be momentarily switched off. Consider what happens to the voltage at the battery terminals. Neglecting the effect of the voltage increment (IR) previously necessary to overcome the internal resistance (this has now vanished since no current is flowing), it will be appreciated that the battery potential will fall along some such line as MN due to the spontaneous relief of the factors out of equilibrium within the battery, factors such as electrolyte non-homogeneity which we discussed earlier. Thus when charging is resumed again the curve continues from N as shown. Had the battery been in use, instead of being charged, the step MN would have gone upwards. Thus clearly the development of any internal lack of equilibrium means that the two voltage curves cannot be coincident, or that the operation is not being carried out reversibly. A reversible process is often described as a sequence of equilibrium states for this reason. More important for the present purpose is the corollary that if a system *is* in internal equilibrium any infinitesimal change taking place in it will possess the characteristics of reversibility.

Reversible Volume Changes of a Gas

The idea of thermodynamic reversibility is such an important one to be clear about that it is worth while to discuss it with reference to two further common examples. The first concerns the compression or expansion of a gas. According to the kinetic theory a gas is to be thought of as composed of very large numbers of individual molecules moving in a chaotic fashion more or less independently of one another. The pressure of the gas is due to the 'bouncing' of these molecules against the walls of the container. If one of the walls happens to be a piston in motion relative to the mass of contained gas the pressure it 'feels' will clearly be greater than that (P) experienced by the stationary walls of the cylinder if the piston is moving towards the gas, and oppositely. The effect is illustrated in Fig. 4.3, and is due to the finite speed

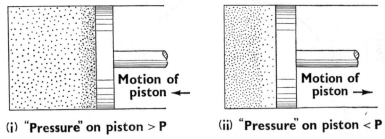

Fig. 4.3. Gas pressure on a moving piston.

with which the molecules react to the piston. Consequently, if the gas is first rapidly compressed and then at once rapidly expanded the two P-V curves will not coincide (Fig. 4.4). They will enclose an area which represents the work lost in the cycle, work which does, in fact, become evident as heat. This effect has nothing to do with the nature of the cycle, i.e. whether the cylinder is thermally insulated (adiabatic process) or in good contact with a thermostat bath (isothermal process). It need hardly be stressed that if the compression and expansion are carried out infinitely slowly (i.e. reversibly) the two curves become coincident, as shown dotted in the figure.

Communication of Heat

The last example concerns the passage of heat. If an object is placed in contact with a source of heat (a body at a higher temperature) it absorbs heat from it and rises in temperature. If the object is to be restored to its original temperature (as the battery was restored to its original charge, or the gas to its original pressure) it will have to be placed in contact with a third body at a lower temperature. When the results of the complete cycle of operations are added up it is found that the hotter source has lost an amount of heat and the colder one has gained it, the object of course finishing at the temperature at which it began. Thus, what has been lost is not work (degraded into heat) as in the two first examples, but heat degraded from a higher temperature to a lower. However, this passage of heat could have

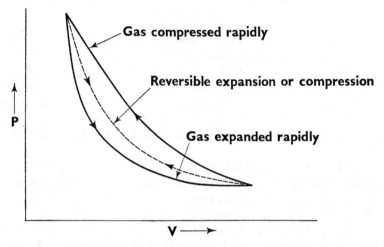

FIG. 4.4. Irreversible volume changes in a gas.

been through a heat engine yielding some work, so it is justifiable to regard even this last example as involving lost work. It hardly needs to be pointed out that if nothing is to be lost the heating and cooling must be done by bodies whose instantaneous temperature differs only infinitesimally from that of the object. And this will mean conducting operations infinitely slowly.

Speed of Reversible Operations

In the foregoing discussion it has often appeared that changes must proceed infinitely slowly to be reversible. While this is true[1] it does not follow that infinitely slow processes are necessarily reversible ones. There are two ways of reducing the speed of a process: either the driving force (the unbalanced e.m.f., the pressure difference, or the temperature differential) can be reduced to a low value, or the opposition can be raised to a high one. The latter can be done by augmenting the electrical resistance, raising the frictional forces between piston and cylinder, or interposing a thermally insulating jacket, to illustrate the three examples. It is only when the speed of the process is reduced by the first means that reversibility is achieved. This is a point that

[1] In *thermodynamic* systems; we are not thinking of *mechanical* systems like a pendulum.

will be readily appreciated. A very slow movement against severe friction is highly irreversible, and so is the imperceptible leakage of heat out of a good Dewar flask. What reversibility requires is the absence of any unbalanced forces (rather than the absence of friction), and it is only when this condition is met that 'infinitely slowly' means 'thermodynamically reversibly'.

Final Remarks

The idea of reversibility belongs essentially to the Second Law. That is why in measuring thermodynamic quantities which depend only on the First Law we do not have to bother about conducting our operations reversibly; the examples of ordinary specific heats or latent heats come readily to mind. However, when the quantities measured involve the Second Law then due regard has to be paid to reversibility when the measurements are made. Here the use of reversible electrodes to measure potential differences is a case in point.

One final word can be added about chemical reversibility. The biologist often meets with chemical reactions which are said to be reversible or irreversible, the conversion of starch into glucose phosphate by phosphorylase being an example of the first, and its conversion into maltose by amylase an example of the second. This use of the words is however quite different from that employed in thermodynamics. A reversible chemical reaction is simply one whose position of equilibrium lies somewhere near the 'middle', so that it is a practical possibility to start at either side of the equation and obtain a useful yield of the substances on the other side. An irreversible reaction on the other hand, is one whose position of equilibrium lies so far to one side that when it has gone to completion virtually none of the original substances are left. Thus, it can never be employed in reverse to generate these substances since the yield would be too little to be of any worth. Whether a reaction is *chemically* reversible or not it can always be allowed to proceed in a *thermodynamically* reversible fashion by suitable control; alternatively it may be allowed to look after itself and waste its potentialities for work irreversibly.

CHAPTER 5

Perfect Gases and Some Other Things

> Pooh began to feel a little more comfortable, because when you are a Bear of Very Little Brain, and you Think of Things, you find sometimes that a Thing which seemed very Thingish inside you is quite different when it gets out into the open and has other people looking at it.
>
> THE HOUSE AT POOH CORNER
>
> *A. A. Milne*

One of the paradoxes of thermodynamics is this: it is a subject developed originally for the very down-to-earth purpose of helping in the design and construction of heat engines, which were needed in large numbers as the industrial revolution gained momentum. Its conclusions were required therefore to be eminently practical ones and not mere exercises in abstract thinking; and yet, in spite of this, one is struck by the fact that historically it made use of idealized substances like perfect gases, and idealized processes like those of frictionless machines operating in reversible cycles, which in actual practice were totally unobtainable. How does it come about that arguments drawn on these lines lead to relations which are found to be precise and exact in the everyday world and not merely approximations to the situations that have to be faced?

This difficulty is likely to be felt particularly by the type of reader for whom this book is written; one whose inclinations as a plant or animal physiologist do not lead him directly into the most abstract realms of thought. It may therefore be worthwhile to spend a few moments thinking about it.

The answer to the problem has in fact already been hinted at in connection with the discussion of reversible processes in Chapter 4. It turns upon this consideration: while a process cannot be conducted quite reversibly in the laboratory there seems to be no relevant reason why reversibility should not be approached as closely as we wish without encountering an insuperable limit. Of course, the cost of human labour and of materials, the limitations of technique and of time do set bounds, but most readers would acknowledge at once that these are irrelevant; they are not as it were 'built in' to the structure of that aspect of the real world which is being considered. Since we have reason to believe therefore that reversibility can be approximated to as closely as we wish, it is quite justifiable to use the notion in the derivation of thermodynamic relations held to be true of the real world.

A very similar sort of argument applies to the use of perfect gases. As every reader will be aware a perfect gas obeys exactly the relation

$$PV = RT \tag{5.1}$$

where P is the pressure of the gas, V the volume of unit mass (a gram molecule or mole), R the 'gas constant' and T the absolute temperature. Absolute temperature has not been specifically defined at this point but this will be returned to later; what is of immediate concern is that equation (5.1) is used to derive thermodynamic results which are expected to be precisely true, although it is perfectly well known that the gas does not exist which obeys it exactly. This difficulty is avoided however by the same sort of argument as was used before: while a perfect gas cannot be obtained with which to charge heat engines, or fill gas thermometers, it is possible in fact, using *any* gas, to approximate as closely to the ideal as is desired. The simple expedient of working at a low enough pressure is all that is needed. That this is a valid way of looking at things is learnt from experiment. Thus a practical laboratory determination of great importance to thermodynamics is that of the absolute temperature in centigrade degrees, on the gas scale, of ordinary melting ice. In outline this is done as follows. A suitably shaped and sized container is filled with *any* gas (some are of course better than others, but only from the practical point of view). Keeping this container in a bath at the temperature to be measured the volume of the gas is varied and simultaneous measurements are made of its volume and pressure, in each case calculating the product PV. This involves only ordinary straightforward operations. Next the product PV is plotted against the volume (V). From experiment it is found that whatever gas is used, the product PV approaches asymptotically to a constant value as the pressure falls (or the volume rises), and Fig. 5.1 illustrates this. But this is by no means all. If the experiment is performed at two different temperatures (say those of melting ice and boiling water), and if the asymptotic values of PV are referred to as A_0 and A_{100}, then the ratio A_{100}/A_0 turns out to be not only independent of the amount of gas we originally took and of the shape and material of the containing vessel, but of the chemical nature of the gas as well; oxygen, nitrogen, hydrogen and even easily liquefiable gases like chloroform vapour will all, experiment reveals, give the same answer. It is in fact the very answer that would be obtained had a perfect gas been available (this last is not a difficult matter to prove, and it is certainly an easy one to appreciate). Knowing the ratio of the two temperatures on the perfect gas scale, and their difference having been fixed by convention as 100°, we have attained our goal. Consequently, if the scale of temperature is defined on the basis of a perfect gas thermometer we are not building on an unreal or at least on an impracticable foundation; we have a perfectly feasible method of determining actual temperatures on it. A perfect gas, in fact, becomes an entity to which we can approximate as closely as we desire. As a constituent in thermodynamic arguments therefore, it will help us to valid conclusions about the real world of, *inter alia*, imperfect gases.

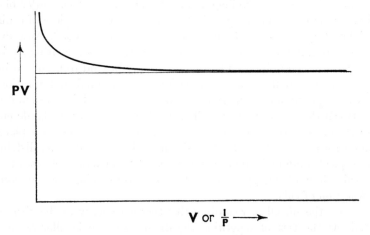

Fig. 5.1. Changes in PV at constant temperature for a real gas.

Thermal Energy

Some of the properties of perfect gases must now be considered which will help us in later discussions, the reason why this class of substance is chosen being simply that it is the simplest. It should perhaps be emphasized that while use is made of perfect gases in developing, for instance, the ideas of absolute temperature and entropy, this procedure is not by any means necessary. Entropy and absolute temperature can be established in quite other ways and the man who is interested in theoretical elegance and rigour would much prefer that these were followed. But this book has in mind a different class of reader, and its object is to help the practically-minded physiologist to grasp difficult conceptions. For this reason a line of development will be followed which is neither historical nor rigorous, but calculated rather to make the subject easily intelligible.

From the molecular point of view a perfect gas can be conceived as consisting of molecules moving about at high speed but interfering with one another, neither by exerting mutual attractions or repulsions, nor by frequently colliding, these two requirements being adequately guaranteed in practice by working with a rarefied, but real gas. If the gas has occupied its container for more than a fraction of a second it will ordinarily have attained that internal equilibrium which consists in the establishment of the random chaotic pattern of molecular movements, a pre-requisite of being able to say meaningfully that it has a definite, unambiguous temperature and pressure. This randomization of the molecular movements is, as was mentioned earlier, an extremely rapid process, since at ordinary temperatures a gas molecule will be travelling at an average speed of the order of one-third of a mile per second, equivalent to passage across a piece of ordinary sized laboratory apparatus (five centimetres) in 10^{-4} second. Now during this process of randomization, and in fact after it, energy is continually being exchanged between individual molecules. It may be said to be stored in the molecules in different modes—translational, vibrational and rotational—and it can be pictured as being rapidly shuffled about both between molecules and between these various modes. The sum total of this very labile energy within the gas is spoken of as its thermal energy. When the gas is placed in contact with a hotter body heat energy flows into it and being distributed in the typical random fashion among its constituent molecules goes to augment this thermal energy, a rise of temperature being the macroscopic manifestation of this.

However, the pitfall of regarding this thermal energy as 'heat' must be avoided. At the risk of repetition the distinction can be illustrated by a simple example. Suppose a sample of gas consists of a mixture of two species in chemical equilibrium,

$$A_2 \rightleftharpoons 2A.$$

We may think, if we like, of the dissociation and association of nitrogen peroxide. The reaction one way will in general be exothermic, that is it will evolve heat; consequently, if the temperature is raised the equilibrium will be driven in the reverse direction, by Le Chatelier's Principle. If the gas mixture is heated the temperature will rise concomitantly with the rise in thermal energy, the molecular motions becoming more vigorous; but in addition, since the reaction moves in the endothermic direction some of the *heat* which enters is, as it were, 'fixed' chemically and does not appear as an increment in the *thermal* energy.

This simple example therefore shows that when a system rises in temperature as a consequence of being heated, its increase in thermal energy is not the same as the amount of heat it has taken in; in fact, the two can be widely different. They are only the same when the heating process is accompanied by no observable change in any of the thermodynamic properties except temperature; no change in composition must occur, nor change in phase, nor significant change in volume, to name but a few possibilities. It is in these restricted circumstances that the notion of an 'absorption of heat' appears to the ordinary reader as something so simple as to need no elaboration. It is just equal to the increase in thermal energy, which in turn can be intuitively appreciated as the energy of molecular Brownian movement. But in fact changes of the sort mentioned (chemical changes, changes of state and so on) are very usual when a system is raised in temperature, and when these occur the notion of heat absorption loses its deceptive simplicity and becomes quite difficult to define. This is one of the reasons why the Danish physicist Brønsted[1] sought to dispense with it; in his approach to thermodynamics two rather different principles replace the well known First and Second Laws, to which however they are together equivalent.

Internal Energy of a Perfect Gas

Returning now to the discussion of perfect gases, it was seen that a perfect gas obeys exactly the relation

$$PV = RT \text{ (per mole)} \tag{5.2}$$

where P and V are measured in the ordinary way and T is measured in the manner already outlined. Perfect gases thus obey at all times a relation which real gases obey precisely only at very low pressures. Now it can be shown very easily to be a thermodynamic consequence of this (though it will not be done here) that the internal energy of a perfect gas depends only on its temperature; in other words, at a given temperature a quantity of perfect gas possesses an internal energy which does not depend on its volume or pressure. In the symbolism of partial differential coefficients,

$$\left(\frac{\partial U}{\partial V}\right)_T \text{ and } \left(\frac{\partial U}{\partial P}\right)_T = \text{zero} \quad \text{(perfect gas).} \tag{5.3}$$

This result is an important one, and it can be appreciated from two angles. Experimentally it is found that if a real gas is allowed to expand into a vacuum it undergoes only a very small change of temperature, and this change becomes closer to zero as the initial pressure of the gas is made less. Expansion into a vacuum is a process in which no *external* work is done or received; in fact it is often said loosely that the gas is made to do work on itself, the fall in pressure being associated with macroscopic movement in the gas which then becomes dissipated as thermal energy as the gas reaches the walls of the containing vessel. Of course in an experiment of this sort it goes without saying that we neither supply nor remove heat; consequently since the gas exchanges neither heat nor work its energy must remain constant. Thus at a low pressure (or for a perfect gas) constancy of temperature is associated with zero change in the internal energy.

The molecular picture of a gas illustrates this property in a rather different way. Consider two equal amounts of the same gas, both at the same temperature but occupying different volumes. Since they are at the same temperature

[1] For full reference see footnote three p. 13.

the intensities of their molecular Brownian movement are the same,[1] and therefore the thermal energies associated with this. If, on the other hand, their molecules attracted each other, work would have to be done against the forces of attraction when the volume increased, and this would represent an increase in the potential energy of the gas with the higher volume (or a decrease, if their molecules repelled); but, since in a perfect gas there are neither attractions nor repulsions between the molecules, there is no such potential energy component dependent on the volume, and the result therefore follows that the energy depends only on the temperature.

Work Done in Isothermal Expansion of a Gas

It is a very simple matter to calculate the work done when a perfect gas expands isothermally, that is, without changing its temperature. The reader will hardly need to be reminded that this condition will only hold if the gas is supplied with heat, otherwise on doing work it will have to call on its own reserves of energy and its temperature will fall.

Imagine a mole of perfect gas enclosed in a frictionless cylinder and piston, the cross-sectional area being A (Fig. 5.2).

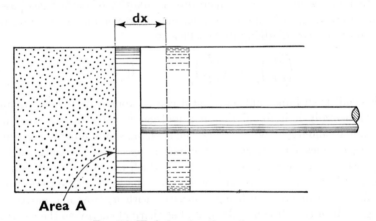

Fig. 5.2. Work done in volume changes.

When the piston is stationary the force exerted on it by the gas will be PA, but as soon as the piston begins to move the force rises above this (if the gas is being compressed) or falls below it (if the gas is being expanded), the effect as between piston and gas molecules being similar to that between an oncoming ball and a tennis racquet. Consequently it must be specified that

---

[1] This is a property of temperature which has not hitherto been mentioned. Its proof depends on statistical theory.

the expansion is carried out *reversibly*, or at infinitesimal speed, in contexts such as the present one.

Under these conditions if the piston is allowed to move a distance dx the work done by the gas will be

$$dW = PA \cdot dx = P \cdot A dx = P \cdot dV \tag{5.4}$$

where dV is the volume swept out by the piston. This formula is incidentally a very important one for the increment of work done by a pressure; it enables

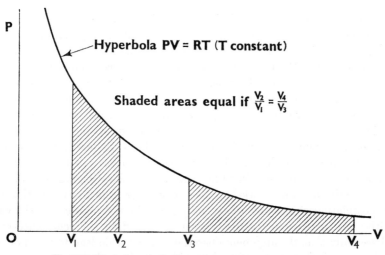

FIG. 5.3. Work done in isothermal expansion of a perfect gas.

us to dispense henceforth with the idea of working in a piston and cylinder. The total work W done in expanding from a volume V_1 to a volume V_2 is found by summing up all such elements as PdV:

$$W = \int_{V_1}^{V_2} P dV. \tag{5.5}$$

Remembering that $PV = RT$, this result is found by the ordinary laws of integration as

$$W = \int_{V_1}^{V_2} \frac{RT}{V} \cdot dV = RT \ln \frac{V_2}{V_1} = RT \ln \frac{P_1}{P_2} \quad \text{per mole.} \tag{5.6}$$

That the result depends only on the ratio of the two volumes is illustrated in Fig. 5.3, where the work done is represented by the shaded areas.

Since the osmotic potential Π of an ideal solution obeys an equation analogous to the gas equation, that is

$$\Pi = RTc \tag{5.7}$$

where c is the solute concentration[1] in moles per unit volume, the expression for osmotic work is similar to the one just derived, namely

$$W = RT \ln \frac{c_1}{c_2} \qquad (5.8)$$

per mole of solute. These two results will be useful in a later chapter.

Radiation

The thermodynamics of radiation is a subject that will be treated a little more fully later, but it is mentioned here because it raises perplexities similar to those belonging to the idea of a perfect gas and frictionless machinery, only rather worse. Radiation obviously possesses energy and it can be shown to exert a pressure. It can be reflected from a *polished* surface in rather the same way as billiard balls are bounced off the edge of a billiard table, and from a *white* surface much as a jet of water is splashed off a rough wall. In view of these analogies with matter it is not surprising to find thermodynamic arguments about radiation which utilize cylinders and pistons with perfectly reflecting walls, and in which the radiation can be subjected to expansion and compression just like a gas. The justification for such concepts is that firstly, practical workshop experience does seem to suggest that it is not postulating things inherently impossible in the physical world when entities are invoked like surfaces with one hundred per cent reflectivity; and secondly that the arguments based on their use do lead to results borne out by experiment, such as the law which predicts that the radiation intensity from a hot body will be proportional to the fourth power of its absolute temperature. However, the further development of this theme will be better left to a subsequent chapter.

---

[1] The analogy is obvious when it is remembered that the concentration is the reciprocal of the volume per mole.

CHAPTER 6

The Second Law of Thermodynamics

> Eeyore was very glad to stop thinking for a little, in order to say 'How do you do?' in a gloomy manner to him.
> 'And how are you?' said Winnie-the-Pooh.
> Eeyore shook his head from side to side.
> 'Not very how', he said. 'I don't seem to have felt at all how for a long time.'
>
> WINNIE-THE-POOH
> *A. A. Milne*

We must now discuss what is one of the most famous of all laws of nature, and one about which a great deal has been written, philosophical, mathematical and prosaic. Its discussion abounds in pitfalls and it is not easy to avoid logical inconsistencies and inadequacies. The treatment we shall attempt here will certainly not be rigorous; the aim will be rather to make it helpful without over-simplification.

The First Law of Thermodynamics is suggested by the fact that once 'energy' is thought of as a constituent of nature this leads us to postulate that it is capable of changing its form from one manifestation to another, and that in these changes it somehow remains constant in amount. It almost comes to be thought of as a sort of 'stuff' which can neither be created nor destroyed, but is for ever fixed in its total amount, at least within the relatively short time span covered by human scientific experience. And it is found that this idea, as a description of our experience of nature, does work. It always leads, so far as can be seen, to correct results, and so it is given the status of an established law.

The Second Law of Thermodynamics takes its rise from quite a different aspect of natural happenings. Perhaps it may be said that this belongs, at the deepest level in our human consciousness, to the realization that there is a certain irrevocability about things which is quite universal in the real world, as opposed to any world of imagination. An action once done or even a thought once entertained, can never be recalled. It may be possible to modify or cancel some of its effects; but the exact *status quo* before it took place can never be recovered. The happening has left its impression on history for all time.

Now what is true in this grand sense is also obviously true in the much narrower sphere of purely physical events, using this designation to embrace all that is covered by the terms matter and energy. Every happening in this sphere—such as the dissolving of a lump of salt, the combustion of a piece of coal, or the evaporation of a dish of water—every such event has about it the mark of irrevocability. It is not of course meant by this that it is not possible, in a sense, to get back to the original situation. It is a very simple matter in fact to recover salt by evaporating its solution, and water by condensing vapour, and similarly for the other examples; but as soon as this is considered we become conscious that such reinstatement of things has been at a price. Without defining too closely at this stage what this means it is obvious that some of our resources have had to be expended to accomplish it, and an unpleasant reminder of this occurs when the gas or electricity accounts are rendered. If the situation is put a little differently it may be described like this. When salt is thrown into water, or a dish of liquid is left uncovered, a process takes place spontaneously—that is, without any help from us or from any other agency—which it would cost the expenditure of effort to reverse. The process which goes of its own accord in one direction will never of its own accord go in the other. If we wish it to go in the other it must be driven, and that means that changes are promoted elsewhere; and these changes, in their turn, can never be expunged except by still others. In fact, a spontaneous process, once it has taken place, has left an indelible mark on the physical world.

Of course in speaking of a process as spontaneous all the relevant information must be included in the specification. It is only under certain conditions, for instance, that an open dish of liquid evaporates spontaneously. If the liquid is a strong aqueous solution, and the atmosphere is very humid, there may not be spontaneous evaporation but spontaneous condensation into the dish. To say that the process will only go of its own accord in one direction, therefore, does mean that all the relevant details have been included in the description of the process, in this case the humidity of the air, and the strength of the liquid solution. Provided we are always precise in this sense the fundamental statement is universally true: all happenings in physical nature involve a degree of irrevocability. They take place spontaneously in one direction but not in the other, and once they have occurred, something has been lost from the sum total of things, which can never be recovered. They involve, in other words, the element of *thermodynamic irreversibility*.

Now it is this very broad and fundamental situation of which the Second Law takes cognisance. The very breadth of its foundation seems to indicate beforehand that it will lead to very important and far-reaching results and in this it does not disappoint us. However, before beginning to try to develop its consequences there are some preliminary considerations that must first be advanced. These will lead to several statements of the Second Law which will serve as starting-off points for its development.

6. THE SECOND LAW OF THERMODYNAMICS

Preliminary Considerations

When a spontaneous change has taken place in a system, it was seen that the *status quo* (so far as that system is concerned) can be restored provided a suitable effort is expended or, to be more precise, provided work is done on it. But from where is this work to come? The answer is that it is obtained from another spontaneous process, and by allowing this second process to go in the direction it wants to it may be 'harnessed' to drive the first one backwards. An instance of this is the dissolving of a metal in acid. This change can be driven in reverse, and the metal and acid recovered by electrolysis, by harnessing a second chemical process in the form of a suitable primary battery, and allowing this second process to take place in its own spontaneous direction. Of course, the harnessed process need not be similar in kind to the first; it could be the fall of water through a turbo generator, or the flow of heat through a thermoelectric device. This line of thought leads to the suggestion that perhaps any spontaneous process could be harnessed in this way, to do work; and growing experience of physical nature confirms that this suggestion is in fact true. With some natural processes this does not seem to be a very startling discovery, but with others it is. Some changes are very obviously connected with energy of one form or another: the release of a spring, or the chemical combination of petrol vapour and oxygen, are examples of occurrences which very clearly give out energy. It is hardly surprising that spontaneous processes of *this* sort can be harnessed to do work, but there are other processes which indeed 'go by themselves', but which seem to involve no energy at all. Thus, if two perfect gases are confined side by side in flimsy containers they will inter-diffuse spontaneously when the separating walls are pierced. No energy is manifested, but there is no doubt that the mixing goes naturally of its own accord. Less exact but more familiar is the case of what happens when a lump of sugar is placed at the bottom of a glass of water. Solution and diffusion throughout the volume is a spontaneous process; but very little energy is involved, the little that is, being principally manifested as a slight change in temperature. Can this type of spontaneous change be harnessed to do work? The answer to this again is given by observation and experience; it is found in every case considered that it can. In the present examples the processes of mixing can be made to do work by using a semi-permeable membrane. For instance, the sugar is confined behind such a membrane fabricated in the shape of a piston in a cylinder (Fig. 6.1). Every botanist knows that under these conditions a pressure develops in the imprisoned sucrose solution, and by utilizing this the system can be made to do work as mixing takes place. Of course the work can only be provided at the expense of some form of actual energy; since no energy worth speaking of is evolved by the process itself, it must come from the only other possible source, the thermal energy of the system. Correspondingly, it is found that when the system is harnessed in this way it actually cools itself, whereas when mixing is allowed to take place unharnessed no such temperature change occurs, but this is an aside. The conclusion to be emphasized is that *all* changes which take place spontaneously in nature can be harnessed to do work, no matter how the energy equivalent of this work has to be provided. It is an interesting and valuable exercise for the reader to think out mechanisms by which various

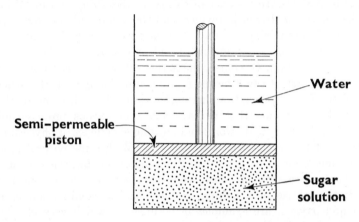

Fig. 6.1. The deriving of work osmotically from the equalization of concentration differences.

types of spontaneous process can be harnessed in this way, examples for this purpose being the equalization of temperature in a red-hot poker, the hydrolysis of starch in the presence of amylase, and the evaporation of a liquid into dry air. Needless to say, the discovery of a way in which a change in the spontaneous direction can be harnessed to do work will provide the answer to the cognate problem of how such a change can be driven backwards by the performance of work.

Lewis's Statement of the Second Law

We are now in a position to appreciate the statement of the Second Law given by a great leader of thermodynamics in America, G. N. Lewis. This is as follows:

'Every spontaneous process is capable of doing work; to reverse such a process requires the performance of work.'

Of course, it must be remembered that here, as always in thermodynamics, words are being used in a very precise sense. Evaporation is a process, in a general use of the word, and so is its opposite, condensation. Both occur spontaneously in nature with equal frequency. What this statement of the Second Law means is that in circumstances in which one occurs spontaneously the other must be 'driven', and that whichever happens to be the spontaneous one in these circumstances can be harnessed to do work. There will be other circumstances when the one now needing to be driven will itself be the spontaneous process, and in those other circumstances it can in its turn, be harnessed.

Lewis's statement of the Second Law is not usually used as the starting point for its development, but it has been introduced here because it does give a different 'slant' on this great principle, and it will be referred to later.

Friction

A slightly different approach to this problem of irreversibility must now be considered. One of the phenomena which by reason of its widespread nature and the impact which it makes on everyday affairs suggests itself as almost the most typical example of irrevocability in physical nature is that of friction. Very great pains are taken to reduce it as much as possible by oiling machinery and designing elaborate bearings which inherently offer as little resistance to relative motion as possible. This is done because it is realized that friction involves the loss of useful work. Once the First Law has been accepted this lost work is thought of as having been converted into heat, and it is almost instinctive to hold the view that this conversion of work into heat is irreversible. Were it not so the problem of friction would not be the serious matter it is; the heat could just be reconverted to work and the losses recovered. That heat cannot be converted into work without loss in this way (as other forms of energy can be converted mutually into each other) is a conclusion forced upon us from the inability of mankind, hammered home by centuries of trying, to construct a perpetual motion machine. One way, in thought, of making such a machine would be by continuously creating energy to make up losses; and the failure to do this leads to the First Law. But another possibility would be to tap off the kinetic energy degraded to heat, transform it to work, and feed it back to the moving parts. The failure to produce such 'perpetual motion of the second order' (clearly quite a distinct possibility from the previous one) leads directly to the Second Law.

Before stating the Second Law in these terms it may be useful to remark that while the degradation of work into heat may be readily accepted as the irreversible process *par excellence*, there are many other irreversible processes in nature, such as the uncontrolled diffusion we discussed earlier. In fact every spontaneous process involves some degree of irreversibility, and so has a right to be included here. Now according to Lewis's statement all spontaneous changes are capable of doing work, and the element of irreversibility comes in when they are used to do less work than they are capable of. Further, when they are made to do work (as seen in the osmotic example) they do it, in part at least, at the expense of heat energy. Here again the irreversible component means that less heat is drawn upon than the process is capable of converting into work, and at the end more heat is 'left over' in the system, as it were, than would have been the case had the change been fully harnessed. Thus in effect it is not too far fetched to say that, even in these other quite different types of irreversible processes, work is lost and heat energy gained.

Kelvin's Statement of the Second Law

A form of the Second Law due to Lord Kelvin can be expressed as follows:

> 'It is impossible to devise an engine which working in a cycle, shall produce no effect other than the extraction of heat from a reservoir and the performance of an equal amount of mechanical work.'

This states very clearly the principle that heat cannot be simply converted into work unless other changes, necessarily of a type which would occur spontaneously, also take place to 'pay' for the conversion. The words 'in a cycle' are added to ensure that the engine is purely an auxiliary piece of apparatus which undergoes no permanent change itself, and which therefore for thermodynamic purposes can be forgotten. Of course, the statement does not mean that as ordinarily understood, heat energy cannot be converted into work; this conversion, loosely speaking, is a commonplace in our industrial society. But it does mean that this upgrading of heat energy, if the phrase may be used, can never be made spontaneous, whatever sort of mechanism is pressed into service. The degradation of work into heat is known to occur only too readily; Lewis's statement of the Second Law implies that its reversal (which we are considering) requires the performance of work; and this work can, again by Lewis's statement, be provided by another spontaneous process. Thus the two statements of the Second Law are seen to be in agreement. In the case of a heat engine what happens can be regarded in the following way: a quantity of heat Q' is upgraded into work, equal to it in amount; and a second quantity of heat Q'' is taken in at a high temperature and rejected at a lower one. It is this second process, obviously in the spontaneous direction, that pays for the first, and it explains why a heat engine always needs to have at least two different temperatures available to it.

The Measure of Change

We must now digress for a moment to consider a problem which may already have occurred to the reader whilst discussing processes and transformations of different kinds; that is, associated with different sorts of physical phenomena. It is this: when a system is transformed from a state (1) to a state (2) it has by definition undergone a change; is it possible to name any general measure of the extent of this change? In other words, is there any general measure which can be used to compare the amount of change in one system with the amount of a *quite different sort* of change in another system? If the two systems undergo the same kind of alteration the matter is obviously a very simple one; but can anything be done when they do not? Can the change in a system undergoing diffusion, for instance, be compared with that in a system reacting chemically, or expanding in volume? It may be of course that while changes can be broadly recognized as being 'more' or 'less' in extent, no precise measure will prove possible. After all, things are known which can be only roughly classified in a quantitative way, like aesthetic merit; so there is no definite guarantee that a precise measure will prove possible in the present case. However, in dealing with the objective world of science and not the subjective one of aesthetics, there is at least a promise that we may be successful.

Consider for simplicity a system which changes between two states (1) and (2) at the same temperature. For definiteness, it might consist of known amounts of sucrose and water at atmospheric temperature and pressure; in pure form in state (1), as a solution in state (2) (Fig. 6.2). In the direction

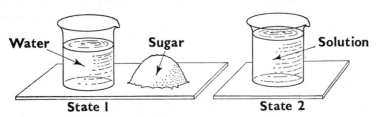

Fig. 6.2. Different states of a closed system.

(1) to (2) the change, which is simply one of solution and diffusion, is a spontaneous one; by Lewis's statement it could therefore be harnessed to do work. Of course, its potentialities in this respect could all be allowed to run to waste. There is therefore no *lower* limit to the work which might be obtained; but there is obviously an *upper* limit, and from what we have said in Chapter 4 it will be clear that this upper limit is realized when the process is allowed to proceed under strictest control, i.e. reversibly. Now consider the suggestion: could not this maximum amount of work be taken as a measure of the amount of change? It is obviously very suitable in many ways, for it would be in units common to all sorts and types of change (chemical, electrical, phase and so on); and it would have the valuable property that two quite

different examples of change which had the same measure could be coupled together and either used to drive the other the exact amount in reverse. For example, a metabolic change could be linked to a transport process and the two would be exactly 'equivalent' if they had the same measure on this scale.

However, while the measure discussed is a valuable one (it is in fact the 'free energy change') it does not quite satisfy the requirements. For one thing it is limited to cases in which the temperature does not alter; where the temperature does alter the measure fails because there *is* no maximum amount of work obtainable in the transition from one state to another (see Appendix II, Chapter 7). But there is a deeper reason still that makes it necessary to look for another measure. The doing of work reversibly does not cover that aspect of change which is to be measured. The swinging of a pendulum involves change, but it is not the sort of thing that leaves a permanent impression on Nature. The pendulum can go on moving back and forth for ever, barring friction. On the other hand once the mechanical energy has been converted to heat something has occurred the effect of which will persist in one way or another for all time. It is this aspect of change which we desire to measure. Suppose therefore, following this clue, that rather than considering the amount of work which a transition (state (1) to state (2)) could do, we consider the heat which it could reconvert into work. This would have the advantage that not only would the universal currency of heat and work be employed to measure diverse types of physical and chemical transitions, but also that the measure would be linked to the *irrevocable* aspect of things.

This at once brings the solution of the problem nearer, but there is still a difficulty. It is connected with what can loosely be called the 'quality' of the heat energy involved. One very obvious way in which heat energy is inferior to work is this: a system which can do work can communicate energy to any other system whatsoever[1], whereas a system which has only heat energy to offer can only impart it to other systems whose temperature is lower than its own. That is why a small torch cell can vaporize tungsten, whereas the largest of gas flames cannot; the torch cell can do electrical work on the tungsten, and there is no theoretical upper limit to the temperature it can achieve. Heat energy is, however, a sort of 'soft currency' in the energy market. By the same token high-temperature heat is less restricted in its usefulness than low-temperature heat; and the upgrading of high-temperature heat into work is a less exacting task than the upgrading of low-temperature heat. This fact ought to figure in the measure of change. Somehow there must be

[1] It might be objected that a battery cannot do electrical work on another battery with an equal or greater e.m.f. In fact it can; all that is needed is an ideal motor-generator to raise its voltage. This sort of expedient applies to all forms of work, but not to heat.

6. THE SECOND LAW OF THERMODYNAMICS

incorporated into the measure some recognition of the fact that heat has a value related to its temperature and that the lower the temperature the less valuable is the heat. To put this into other words, the ability to convert a quantity of heat at a low temperature into work (that is, to raise this energy above any temperature limitation) is a bigger thing than to convert the same quantity of heat at a higher temperature. This situation can be most easily accommodated by taking as a measure of the extent of change not the heat Q which it might be used to upgrade into work, but the quantity Q/T. When T is small this obviously weights the measure in the direction required and vice-versa[1].

The argument can now be summarized as follows. When a given system undergoes spontaneously a defined alteration, the process can be harnessed to do work (Lewis); and if it is kept under rigid control this work is a maximum for a given pathway. Part of this work is provided by the conversion of heat energy. Since the opposite process (the degradation of work into heat) is the manifestation, *par excellence*, of the principle of irrevocable, permanent change in Nature, it is natural that this amount of heat energy upgraded should be taken as a measure of the alteration. It is only necessary to modify it by dividing it by the temperature of conversion to make it a really satisfactory measure; this division allows for the fact that to degrade work into low-temperature heat is to carry the process further than to degrade it into high-temperature heat. In symbols, the measure of alteration or change in the system becomes

$$Q_{rev.}/T$$

where $Q_{rev.}$ is the amount of heat (the maximum) which would be absorbed by the system as it is taken through the process of alteration reversibly, and T is its absolute temperature.

The reader who is already familiar with classical thermodynamics will recognize at once that what has been done is merely to arrive at a definition of the entropy change of the system, but in a way quite unrigorous to say the least. Since the idea of entropy [2] will have to be developed more exactly almost at once, what is the purpose, it may be asked, of this long digression? To some readers indeed it may have no purpose; but to others, especially those whose interest lies in the more concrete aspects of science, it may be useful in indicating how it comes about that a measure defined as it is in classical thermodynamics in a way so abstract and so abstruse, should possess such a commonplace meaning as the 'measure of change'. If it has made the notion of entropy, as classically defined, seem a little less hard to appreciate, it will have served its purpose.

[1] A further and most important result of dividing by T will become apparent later.
[2] The word itself comes from the Greek term for 'change'.

Source of the Work Done

Before going on to derive the concept of entropy more formally consider again the point already discussed, namely the origin of the actual energy which balances the work when a spontaneous process is harnessed. Since it is always a help when dealing with abstract ideas to take concrete examples, consider three typical cases: firstly, the expansion of a gas from a high pressure to a lower one; secondly, the complete oxidation of a carbohydrate to carbon dioxide and water; and thirdly the change that takes place in a silver iodide battery on discharge. In the first case it is obviously a simple matter to harness the expansion. All that is needed is a piston and cylinder. Suppose for simplicity that the gas is a perfect one. As it is allowed to expand —very slowly if the operation is to be conducted for maximum yield—the gas does work. On what energy supply has it drawn? Expansion itself is clearly not normally an exothermic process. It could only be such (that is, liberate heat and result in a rise of temperature) if the individual molecules repelled each other (and so tended to speed up when allowed to separate), or if the gas was a mixture which reacted chemically when its volume increased; both of these possibilities are denied by our choice of a perfect gas. So obviously the gas must be doing work at the expense of the only other possible source, its own thermal energy. Accordingly, it is found that under these conditions the temperature of the gas falls, or if it is in close contact with substantial surroundings, that its temperature remains constant while it absorbs heat from the surroundings. This is therefore a case in which the work obtained from a harnessed spontaneous process is all done at the expense of thermal energy. To anticipate a point to be dealt with in a future chapter, in the case where the surroundings maintain its temperature constant, the gas loses *free energy*, but gives up no actual energy at all (a property of a perfect gas being that its internal energy is changed only when its temperature is changed). The contribution the gas itself makes to the work done when it expands isothermally is not therefore to supply part of the energy equivalent of the work; rather it is to suffer on expansion a sort of degradation (an increase in entropy) which pays for the simultaneous upgrading of some heat energy from the surroundings. It is this latter energy which is the equivalent of the work done, and which brings the overall process into line with the First Law.

Now consider the second example. This differs from the first in that the process itself (one of chemical oxidation) releases a large amount of energy. When allowed to run to waste unharnessed in a suitable calorimeter this is manifested as heat, and we say that the process is highly exothermic. However, if the oxidation is conducted under strict control (i.e. reversibly) this chemical potential energy can be fully drawn upon to provide work (including in this term the conversion to other alternative forms of chemical energy). This is in fact what the mitochondrion does when it conducts the process of oxidative phosphorylation. At this point an interesting question arises. If this harnessing were carried out in a perfectly reversible manner, would any tendency be found for the system to change in temperature as in the previous case? In other words, would there be any tendency for the heat energy of the system to be drawn upon (or oppositely) as well, in the interest of doing work? The answer is that in this particular case there would be very little. Only a small amount of heat energy would be involved, and almost all of the work done in the completely harnessed oxidation would be done at the expense, not of heat energy, but of the potential energy of molecular configuration released by the chemical process itself. But both sources would be involved to *some* extent. In the light of the previous discussion it will be seen that here then is a case where the process is associated with very little increase (or decrease) of entropy; for it is this factor, it will be remembered, which 'pays' for the upgrading of heat energy into work.

The third case can be dismissed quickly, for it is intermediate between the other two. Of the work done by the battery on very slow discharge roughly 94% represents a transformation of chemical potential energy, and the other 6% represents heat energy which has been upgraded. Compensation for the latter component is of course required, and there is a small degradation of the system in the sense of an entropy increase.

To sum up: when a spontaneous process is harnessed to do work this work is done at the expense of two sources of actual energy. Firstly, the process itself may release potential energy, and secondly, heat energy may be drawn on. In the second case compensation is required and the system must accordingly suffer some sort of degradation; this is associated with its increase of entropy. These two sources of actual energy may be involved in different proportions, but both are always concerned to some extent. Only in special cases (like perfect gases) is only one source of energy involved.

The Mathematical Derivation of Entropy

There now follows a more rigorous development of the idea of entropy, and this will be linked directly to the notion of absolute temperature on the perfect gas scale. It has been seen that this scale is one which can in actual fact be experimentally realized, and although the following procedure would not satisfy real students of the subject it has the advantage that this notion of absolute temperature is a fairly easy one, and is generally familiar to plant and animal physiologists alike.

Suppose a gram molecule of a perfect gas (the amount being chosen for convenience alone) is confined in a suitable vessel. This will be required to have a volume which can be adjusted as necessary, and again as a matter of pure convenience it can be imagined to take the form of a cylinder and piston. Needless to say these must be frictionless; and if in spite of all that has been said before this seems to divorce the discussion from reality consider instead that the gas is confined in a cylindrical vessel made from a light springy metal bellows. Then at least there will be no question of heavy frictional forces consequent on change of volume. The state of the system (the gram molecule of gas) can be described by the very simple and easily measured parameters P, V and T. These however are not all independent; any two of them can be made to have chosen values, but the third will then be automatically fixed. In fact the three will be related by the equation

$$PV = RT. \tag{6.1}$$

Suppose now that the gas happens to be in the state (1) in which its parameters are P_1, V_1, T_1 and that it is required to bring it to the state (2) in which they are P_2, V_2, T_2. The first point to remember is that this can be done in an infinite number of ways. If attention is directed to its pressure and volume and simultaneous values of these are plotted during the transition it is clear that they can be manipulated so that the condition changes along any chosen line in the P-V diagram (Fig. 6.3).

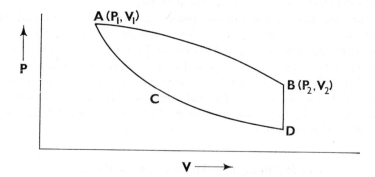

FIG. 6.3. Alternative pathways between two states.

Thus the bellows might be expanded at once from V_1 to V_2, allowing the pressure and temperature to look after themselves. This might involve moving along the line ACD. Then the bellows might be fixed and a bunsen flame held near them till it is observed that the pressure has risen (along DB) to the value required. In the interests of a concrete understanding of the point the reader would do well to think out other ways in which the apparatus might be experimentally manipulated to secure the same final result.

6. THE SECOND LAW OF THERMODYNAMICS

The second point to remember is that however the manipulation is carried out the operator (in more objective and formal language the 'surroundings') has always to be prepared to do two things to the gas: firstly, perform work on it (for example, by pushing in the bellows to a smaller volume); and secondly, provide it with heat from a suitable source. Of course it is understood that the work and heat may be either positive or negative quantities, and in fact in the sequence (ACDB) discussed above the work was negative (that is, it was done not on, but by, the gas) and the heat was positive.

The matter can now be regarded quantitatively. In a particular case suppose that the operator has had to impart a quantity of heat Q to the gas, and has had to do an amount of work W upon it. The sum of these two is of course the increase in internal energy (ΔU) of the gas, as was seen in the discussion of the First Law. Thus

$$\Delta U = Q + W$$

or
$$Q = \Delta U - W. \tag{6.2}$$

If the usual mathematical stratagem of considering infinitesimal changes is now adopted it is found that it is possible to give alternative forms for ΔU and W. If the volume changes from V to $V + dV$ then providing the change is exceedingly small the average pressure operative during the change can be taken as exactly P, and the work done by the gas becomes PdV. The quantity W, however, is the work done *on* the gas, so it must be written $-PdV$. Similarly, if the temperature changes by an amount dT the increase in internal energy is $C_v dT$, where C_v is the specific heat of the gas at constant volume (see Chapter 5). This last relationship is true simply because that is how the specific heat is defined, and its use would not really lead anywhere were it not for the fact that in the case of perfect gases there is further important information about the specific heat which will be introduced in a moment. Therefore for an infinitesimal change:

$$dQ = C_v dT + P dV. \tag{6.3}$$

This equation should be perfectly easy to visualize as a consequence of the First Law. It states merely that the heat supplied must be equal to the sum of the increase in internal energy of the gas and the work done *by* it. In spite of its simplicity, however, it has some very interesting consequences.

Suppose the change undergone by the gas is not an infinitesimal one, but a finite one, perhaps quite large. The quantities appearing in the equation will then be expressed as integrals, and the total heat taken in can be written as

$$Q = \int dQ = \int C_v dT + \int P dV. \tag{6.4}$$

This is all quite straightforward, and the equation still retains the meaning just expressed in words; but if an attempt is made to calculate definite values from it difficulties occur, since whereas the first integral on the right hand side has a well defined value as it stands, the second has not. For a perfect gas the specific heats C_v and C_p depend only on temperature [1] and the first integral can be visualized as the area under the curve relating C_v and T (Fig. 6.4). It does not matter in this connection how P and V have changed

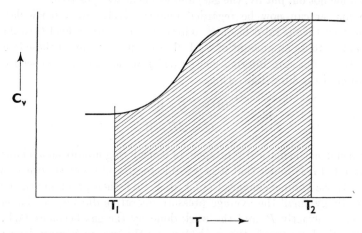

FIG. 6.4. Variation of the specific heat of a perfect gas at constant volume with temperature.

during the transition; a knowledge of the end states fixes the value. In the case of the second integral the value cannot be calculated, unless P can be expressed completely as a function of V. This is a *mathematical* necessity (Chapter 3) and it clearly implies that the whole of the pathway of change has to be known, and not just the two end points. Put in other words the value of the integral ($\int P dV$), and so of Q, will be different for every different path between (P_1, V_1, T_1) and (P_2, V_2, T_2), while the value of the integral ($\int C_v dT$) will always be the same. Thus no very fundamental meaning can be attached to Q in the case under consideration, though it has already been seen that for a perfect gas such as the one used here, there is a fundamental meaning for $\int C_v dT$. It is in fact the increase in internal energy.

An interesting point now arises. Suppose equation (6.3) is divided by the absolute temperature T. This gives the result

$$\frac{dQ}{T} = C_v \cdot \frac{dT}{T} + \frac{PdV}{T} \; . \tag{6.5}$$

---

[1] This is a consequence of the fact that for a given perfect gas the internal energy is a function of temperature alone (see Chapter 5).

6. THE SECOND LAW OF THERMODYNAMICS

Now recall two fundamental properties of a perfect gas. The first is that $PV = RT$, or $\dfrac{P}{T} = \dfrac{R}{V}$. This enables equation (6.5) to be rewritten as

$$\frac{dQ}{T} = \frac{C_v}{T} \cdot dT + \frac{R}{V} \cdot dV. \tag{6.6}$$

The second property is that C_v, and so $\dfrac{C_v}{T}$, is a function of T alone. It is apparent, therefore, that equation (6.5) or (6.6) is immediately integrable without any further information being given as to the actual pathway of change. However many experimenters using different programmes evaluate the quantity $\int \dfrac{dQ}{T}$, the result will be always the same, provided that the work is competently carried out and begins at (P_1, V_1, T_1) and ends at (P_2, V_2, T_2). This constitutes a position rather similar to that which led, on the strength of the First Law, to the recognition that there was such a quantity as U, dependent only on the state of the system and which could be given the significance of a quantity of energy. This followed since changes in U, defined as being equal to $(Q+W)$, proved to be independent of the pathway of change and dependent only on the two end states. Now a quantity $\int \dfrac{dQ}{T}$ has been discovered which, at least for a perfect gas, has a value dependent only on the initial and final states. On the strength of the Second Law it will soon be seen that this result holds not only for perfect gases but for all substances, and for all conceivable types of change; and therefore it is possible to establish another quantity S defined analogously to U by the equation

$$\Delta S = \int \frac{dQ}{T}. \tag{6.7}$$

This quantity S, as the reader will have guessed, is called the entropy of the system, and it plays a part in thermodynamics as important and as fundamental as temperature itself. Even after the earlier discussion on 'change' the concept of entropy will probably seem abstruse and difficult, and until its statistical interpretation has been discussed it will not become much easier. Meanwhile there is a clarifying statement that needs to be made.

In developing the primary equation (6.3) the expression for the work done by the gas was taken as PdV. This is only true if the expansion is carried out *reversibly;* for if a surface is moving away from oncoming gas molecules the force they exert on it when they strike it and bounce back is less than it would be if the surface were stationary. Consequently the piston 'feels' a pressure less than the real pressure of the gas, and the work done is less than PdV. The reverse is of course true if the gas is being compressed. Thus in the expression for the entropy change it is better to be explicit and specify that the heat taken in must be measured when the chosen path is traversed *reversibly*:

$$\varDelta S = \int \frac{\mathrm{d}Q_{\text{ reversible path}}}{T}. \tag{6.8}$$

Entropy Changes in a Wider Context

The validity of the concept of entropy (that is, the conclusion that the quantity $\int \frac{\mathrm{d}Q_{\text{rev.}}}{T}$ is independent of the path traversed between two given states) has, of course, only been established for the case of volume changes in a perfect gas. That it holds for any sort of substance or any type of change whatsoever is a consequence of the Second Law, and can be demonstrated in outline as follows. Suppose a system consists of two engines; one is built around a frictionless piston and cylinder containing a perfect gas, and the other around a quite arbitrary mixture of substances capable of undergoing any sort of change, conventional or outrageous. For definiteness this second engine may be regarded as having as its essential feature an accumulator battery upon which electrical work can be done, and from which electrical work can then be obtained. The two engines are arranged so that they are coupled together, and either of them can then be used to drive the other in reverse. Normally this will involve some sort of special 'link' to convert the work given out by one into a form of work suitable for absorption by the other, an example of such a link being a dynamo. However, this conversion of one form of work into another can be without loss, and beyond recognizing the necessity and possibility of using such a link this is of no further concern. The engines are operated reversibly in cycles, so that there is also no ultimate concern with any changes that take place inside them, since they can always be left just as they were found. Consequently the only changes which need be taken into account are the interconversion of heat into work. Since one engine is driving the other (that is, doing work upon it) the net work involved in both cycles must be equal; so also must the net heat, since each engine converts either its work input into a heat output, or vice versa. All this follows from the First Law. Now as each engine operates, its working substance will in general go through a series of temperature changes, and by suitably matching the relative sizes of the engines the limits of these changes can be made the same for both engines. One has only to reflect for instance, that if a given amount of electrical energy is passed into a small accumulator battery, the accompanying temperature changes in the battery will be greater than if the same amount of energy is put into a larger battery.

Now consider the very simplest form of cycle, in which heat Q_1 is taken in from the surroundings at a higher temperature T_1 on part of the 'outstroke' and heat Q_2 is rejected at a lower temperature T_2 on part of the 'instroke', the transition between the two temperatures being made by momentarily cutting off the working substance from either gain or loss of heat from its surroundings. Then for the perfect gas engine,

$$\frac{Q_1}{T_1} = \frac{Q_2}{T_2} \tag{6.9}$$

since each represents the entropy change between the fully expanded and the fully compressed conditions (this is of course, operating reversibly). Since it is essential that the reader follows each stage in the argument this conclusion may be illustrated with a diagram. It refers to the engine of the two, whose properties are well known, and it relates the simultaneous changes in pressure and volume undergone by the 'working substance', the sample of perfect gas.

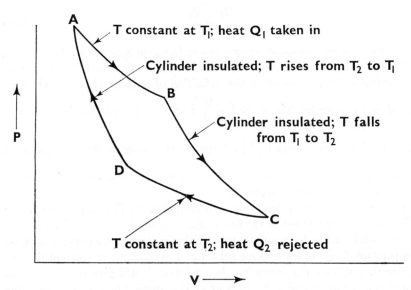

FIG. 6.5. Carnot cycle for a perfect gas.

Starting from A (Fig. 6.5) the gas is allowed to expand very slowly to B in close contact with the surroundings which maintain the temperature constant at T_1. In this process, being a perfect gas, it loses no internal energy, but absorbs an amount of heat Q_1 equivalent to the work it does. At a suitable point B the cylinder is surrounded with an insulating jacket and the gas is allowed to expand further [1] to C. Since it cannot absorb heat now the temperature of the gas falls and the curve becomes steeper than before. A further quantity of work is done, but no more heat is absorbed. At the point C the gas has the volume previously decided upon, and, if the point B has been chosen correctly, the required temperature (T_2) as well. This completes the outstroke. Now A is returned to by the simplest possible alternative path, provided by first an isothermal compression CD in contact with new surroundings at T_2, and secondly by an adiabatic compression DA. If the point D at which the thermally insulating jacket is applied has been correctly chosen the point A will be the final end point, thus completing the cycle. Along the section CD a quantity of heat Q_2 is rejected by the gas, but naturally none is involved along DA. If the entropy change in the gas between the points A and C is considered, the result is clearly either $\dfrac{Q_1}{T_1}$ or $\dfrac{Q_2}{T_2}$ since both of these expressions relate to a reversible path between the same two end points. Consequently we must have

$$\frac{Q_1}{T_1} = \frac{Q_2}{T_2} \tag{6.10}$$

since the entropy change as shown before has a unique value. In passing, a slight confusion in signs must be avoided, since Q_2 has been defined as heat *rejected*. To accommodate the definition of entropy change $-Q_2$ must therefore be used; but the result $\left(-\dfrac{Q_2}{T_2}\right)$ refers to the passage from C to A, and the alternative result $\left(\dfrac{Q_1}{T_1}\right)$ referred to the opposite direction (A to C). Hence the sign must again be changed before the two quantities are set equal; this gives equation (6.10).

We now turn our attention to the second engine. It will be remembered that this has been so matched to the first engine in size, that when driven by it in a cycle with two isothermal and two adiabatic stages, the upper and lower temperature limits of its working substance are the same as those of the first engine (i.e. T_1 and T_2). Since it is being driven it will give out heat at the upper temperature and take in heat at the lower. Let the two quantities of heat involved be Q_1' and Q_2'. The crucial point in the argument is to show that these two quantities are respectively equal to Q_1 and Q_2, no matter what the nature of the reactions undergone by the working substance of the second engine may be (chemical, electrical, nuclear, or just plain volume changes of an imperfect gas). The demonstration is as follows.

---

[1] These two stages are of course 'isothermal' and 'adiabatic' respectively.

Firstly, it must be the case that
$$Q_1 - Q_2 = Q'_1 - Q'_2 \tag{6.11}$$
since, if this relation is not true, the net heat abstracted from the surroundings by the first engine in a single cycle will not be balanced by the net amount returned by the second, and heat energy will have been created or destroyed. It has to be remembered that the only part the surroundings play in the transaction is to supply or remove heat from the engines[1]; and that, since the engines return at the end of the cycle to their original states, changes in the energy stored up within them (i.e. changes in their U's) do not concern us.

Having now established equation (6.11) by appealing to the First Law it is rearranged as
$$Q_1 - Q'_1 = Q_2 - Q'_2 \tag{6.12}$$
and appeal is made to the Second Law. For the equation now says that the heat withdrawn per cycle from that part of the surroundings at the temperature T_1 is balanced by an equal amount given back to that part of the surroundings at the temperature T_2 as the double engine operates; therefore heat disappears continuously from one part of the environment and appears in an equal amount in another part at a different temperature. Since all operations are conducted reversibly, which engine is the driver and which the driven can obviously be chosen to make the transfer of heat an uphill one, that is, from a lower temperature to a higher one. And all this will be at no cost in work, since all the work that one engine needs, the other supplies. This is clearly a result which the Second Law (cf. Clausius's statement below) declares to be impossible. Hence the only conclusion left is that the quantities on each side of equation (6.12) are each zero, in fact that
$$Q_1 = Q'_1$$
and
$$Q_2 = Q'_2. \tag{6.13}$$
From this it follows at once (using equation (6.10)) that for the second engine also, $\dfrac{Q'_1}{T_1} = \dfrac{Q'_2}{T_2}$, and at least for the two reversible paths considered (ABC, ADC, Fig. 6.5) the value of $\displaystyle\int \dfrac{dQ_{\text{rev.}}}{T}$ is the same for any sort of substance (pure or mixed) and for any sort of reaction to which it is exposed. It only remains to show that, for any path whatsoever (traversed reversibly) the integral has the same value, to complete the demonstration. This can be done in outline as follows.

Let Fig. 6.6 represent the 'indicator diagram' for the working substance, or system of substances. In the case of a gas engine it will be a P-V diagram; if an accumulator battery is being considered it might be a diagram relating e.m.f. to the amount of current taken from the cell. For simplicity it is drawn

---

[1] No *work* is involved so far as the surroundings are concerned. The reader should think out the implications of the statement that one engine drives the other.

for the former case, imagining however that an imperfect gas is being dealt with for which the proposition still has to be proved.

Consider any arbitrary pathway ACB between A and B. Through B draw the isothermal BD, and let the space between A and B be divided up by adiabatic lines AD, M_1N_1, M_2N_2 and so on as shown. These adiabatics will cut the line ACB into a number of segments. Let these segments have short isothermals $I_1, I_2, I_3 \ldots$ drawn through their mid-points to cut the two nearest

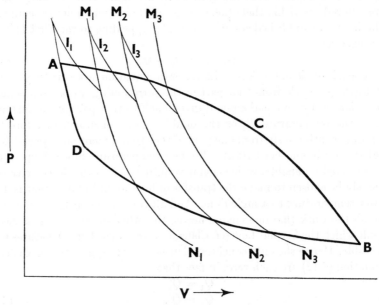

Fig. 6.6. 'Indicator diagram' for an arbitrary system.

adiabatics. Now imagine the working substance to travel, not along the original path ACB but successively along the isothermal lines $I_1, I_2, I_3 \ldots$, passing from one of these to the next up or down a short length of adiabatic. The path will be somewhat as shown in Fig. 6.7, with the original path shown as a heavy line. In evaluating the expression[1] $\sum \dfrac{Q_{\text{rev.}}}{T}$ along such a 'stepped' path we can clearly forget the adiabatic stages, since by definition no heat is exchanged along them. This leaves the isothermal sections ($I_1, I_2, \ldots$), each having its own temperature. The value of $Q_{\text{rev.}}/T$ for each will be the same as that for the corresponding section of the isothermal DB lying between the same two adiabatics (this has already in effect been proved); consequently

---

[1] The symbol $\sum$ is used instead of the integral sign because finite stages are being considered instead of infinitesimals.

the sum total along all the short isothermals $I_1, I_2, I_3 \ldots$ will be the same as the value along the single isothermal DB. Since AD is an adiabatic, this gives the total for the whole path ADB. It only remains to point out that by making the number of subdivisions large enough the 'stepped' pathway (Fig. 6.7) can be made to agree as closely as we please with the actual one. Therefore the

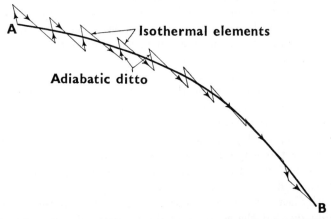

Fig. 6.7. Splitting up of an arbitrary pathway into isothermal and adiabatic stages.

conclusion can be drawn that *whatever* sequence of changes any system goes through in passing from one state to another, provided it does so reversibly, the value of $\int \dfrac{dQ_{\text{rev.}}}{T}$ is determined solely by the initial and final states. This completes the proof that equation (6.8) is a valid one, not only for a perfect gas, but also for any system whatever; and not only for volume changes, but for reactions of the widest and most diverse kinds imaginable[1].

Some Miscellaneous Comments

Before leaving the classical treatment of entropy there are a few miscellaneous comments which may profitably be made. They all follow from the fundamental definition (equation (6.8)).

1. As in the case of the internal energy U, only *changes* in entropy are defined, not the absolute values. This is true at least from the point of view of classical thermodynamics, though other branches of physics may be able to supply further information.

2. As with many other thermodynamic quantities, to speak of the entropy of a system really presupposes that that system is in internal equilibrium or

---

[1] Strictly the proof needs to be developed further for the case where the system has more than the two independent variables considered here.

else its state could not be reached reversibly. In practice it is sufficient if it is nearly so. It will be remembered that the same is true of the ideas of temperature and pressure. Putting a beaker of water into internal motion by stirring (even violently) does not make it nonsense to speak of its temperature; but it is inadmissible to enquire about the temperature of a gas mixture at the moment of explosion. The same sort of thing is true about entropy.

3. Like mass and volume, but unlike temperature, entropy is an extensive property. Doubling the size of a system doubles its entropy, and the total entropy of a complex system is equal to the sum of the entropies of its individual parts.

4. If heat is to pass reversibly into a system from its surroundings, the latter must ideally be at the same temperature as the former, or the heat would go in 'with a rush' and the change (considering the happenings *within* the system itself) could not be perfectly reversible. However, in practice the surroundings can often be appreciably different in temperature from the system under consideration without upsetting the measurements. If for instance some cold solution is placed in a Dewar flask and the whole immersed in a thermostatically controlled bath kept at a higher temperature, the cold solution will naturally heat up, but very slowly. The heat entering the solution leaks in so gradually that the contents of the flask are always substantially in internal equilibrium, with no appreciable temperature gradients developing within them. In such a case, as consideration of the original argument with perfect gases will show, the change in entropy of the solution in the flask is

$$\int \frac{dQ_{rev.}}{T}$$

where T is the temperature of the *solution*, not that of the bath. The requirement that the whole process should be carried out reversibly if the entropy change undergone by the system is to be measured, means that attention must be kept on the system itself during the measurement, to ensure that it never strays more than a minute distance from internal equilibrium; the surroundings can be left to their own devices so long as they do not upset the primary requirement just stated.

FURTHER STATEMENTS OF THE SECOND LAW

The statement of the Second Law due to Lewis, and one based on the ideas of Lord Kelvin have already been quoted. To help further in grasping this very important principle some additional statements are added below.

The first is due to Clausius, who originated the term entropy. It can be put in the form:

> 'It is impossible to construct a device that, operating in a cycle, will produce no effect other than the transfer of heat from a cooler to a hotter body.'

Expressed more simply this says that heat will not of itself flow from a colder to a hotter body.

The second statement is often known as Kelvin's Principle of the Dissipation of Energy. It is obviously fairly close to Lewis's statement:

'Every irreversible process forfeits an amount of work which could be obtained if it were carried out reversibly.'

It is not difficult to appreciate intuitively that these statements are equivalent; but a final statement, also due to Clausius, is one which will be better understood after the chapter on Entropy and Free Energy:

'The entropy of the Universe is continually increasing.'

In this respect the Universe, regarded as an isolated system, is like a wound up clock, subject to a process of becoming run-down. Perhaps, after all, the Second Law has something to do with the quotation which headed this chapter.

CHAPTER 7

Entropy and Free Energy

'Hallo!' said Piglet, 'what are you doing?'
'Hunting,' said Pooh.
'Hunting what?'
'That's just what I ask myself. I ask myself, What?'
'What do you think you'll answer?'
'I shall have to wait until I catch up with it,' said Winnie-the-Pooh.

WINNIE-THE-POOH

A. A. Milne

As it was defined formally in the previous chapter entropy may appear to be a curious and interesting property, but not one which is likely to have much real usefulness in practice. However, this is in fact not the case, and it will be the purpose of the present chapter to discuss some of the wider implications of the concept.

We start by imagining that we have a system (we are, of course, still thinking of a *closed* system) liable to spontaneous change. This qualification does not mean that the system is not in internal equilibrium, since the spontaneous change to which it is liable may be temporarily prevented by a physical or a chemical barrier, as for example with a glass of water and a lump of sugar separated by an air space, or an organic monomer stabilized from polymerization by an anti-catalyst. We suppose that the system is enclosed in a vessel with rigid non-conducting walls, like a very strong Dewar flask. This sort of vessel is chosen to prevent *any* form of gain or loss of energy, the rigidity preventing change of volume (which would imply work) and the non-conducting quality preventing passage of heat. The restraint (such as the anti-catalyst, or air barrier) is next removed and the spontaneous change is allowed to take place. As a result of these manipulations the system undergoes a transformation without exchange of energy (that is, $U_2 = U_1$ or ΔU = zero) and arrives at a new state differing chemically or physically or both from the initial state. Incidentally, in experiments of this sort it sometimes happens that the temperature remains virtually unaltered, as would be the case for the dissolution of the sugar; at other times it rises (or even falls) sharply, as in the case of many polymerizations or oxidations. We suppose next that the experiment is repeated (that is, the *identical*[1] change is promoted) but in a different and non-isolated way, this time harnessing the spontaneous

[1] This implies that the initial and final states are respectively identical in the two experiments (that is, $T'_1 = T''_1$; $T'_2 = T''_2$, and so on for other variables).

process to do some work. This will require that it is conducted not in the original vessel, but in one specially designed, such as a metal cylinder with a semi-permeable piston, an arrangement that would suit the first example given (see Fig. 6.1). Lewis's statement of the Second Law asserts that a positive contribution of work can always be obtained from a process which is spontaneous. Now consider again the important question: where does the actual energy come from to balance the work? The First Law requires that it comes from somewhere; but the system itself gives up no energy, since ΔU for the transformation is zero; this has been arranged for by choosing the terminal states in the light of the first experiment. Further, the surroundings provide no work, since there is no volume change[1]. There is therefore only one answer; the system when harnessed must take in heat from the surroundings and do work at the expense of this heat. But if the system takes in heat when harnessed in this way, the implication is clearly that its entropy has increased (one has only to think of the definition of entropy to realize this); this conclusion applies also to the first experiment, which has the same end states. This leads, therefore, to a point of quite fundamental importance: *in every spontaneous process in an isolated system there is an increase of entropy.* It must be noticed clearly that this statement is only true for an *isolated* system, for if the argument is followed carefully it will be seen that it only holds if ΔU and ΔV are both zero, and this will not be true if the system is free to gain or lose heat, or exchange energy as work, with its surroundings.

The limitation which has just been mentioned, to the use of what may be called the Principle of the Increase of Entropy, is not such a restrictive one as it may seem at first. It is true that most of the systems dealt with by physiologists cannot even approximately be regarded as corked up in Dewar flasks; they are usually in fairly open contact with their surroundings. But that does not mean to say that the principle cannot be applied. Suppose a metabolic change takes place in such a system as a mitochondrion, an organelle in close thermal contact with the general cell fluid. Any heat or other energy released will make its effect felt with decreasing intensity the greater the distance from the scene of action; and if it is sufficiently far away a closed surface drawn around the organelle can be imagined across which the flow is virtually nothing. Then it is only necessary to consider all the changes which take place within this surface to be able to apply the Entropy Principle, for it clearly makes no difference to the original reaction whether this surface is, as supposed, an imaginary one, or on the contrary an actual rigid insulating wall.

The Principle of the continual Increase of Entropy is in fact the most general principle of its kind which thermodynamics gives us, since it applies to the sum total of all physical things, the Universe itself, in so far as it can

[1] Even with the piston and cylinder arrangement. The reader should think this out.

be considered as a closed system. However, while it can always in principle be applied it is not really very convenient for the physiologist; to have to complicate calculations in the way suggested by including in them a part at least of the surroundings is in fact most inconvenient. Fortunately thermodynamics comes to his assistance by pointing out another property of the system which, from his point of view, behaves in a way much more amenable to his requirements. But this is anticipating. Let us see if a more suitable criterion can be found for the physiologist's use than this fact of the spontaneous increase of entropy. First of all consider what the conditions are which the criterion must accommodate, by thinking of a concrete case. The physiologist knows that in the cell there are all the ingredients present to form a certain chemical substance, say glucose phosphate from glucose and phosphoric acid. Given the right enzyme, he might enquire, would the synthesis take place spontaneously? Now clearly, whichever direction happens to be the one in which the reaction goes of itself, it will take place in the cell without substantial change of temperature; the surrounding bulk of the plant or animal body sees to that. The physiologist therefore needs to compare the materials appearing on the two sides of his equation with respect to some property (which has yet to be discovered) measured *at the same temperature* for both, if he wishes to be able to insert an arrow into the equation. That entropy by itself is no guide here can be illustrated with reference to the previous example of polymerization. The monomer polymerizes in the Dewar flask with a rise of temperature and (as indicated by the spontaneity) an increase of entropy. To reduce it to its original temperature heat must subsequently be abstracted, and this implies a lowering of its entropy. This final lowering of entropy bears no necessary relation to the initial rise; so the entropy of the polymerized block when the temperature has been restored may be either greater or less than the entropy of the original monomer. Thus a knowledge of the relative entropies of the monomer and polymer *at the same temperature* is not enough to tell how the arrow is to be placed in the chemical equation connecting them. To do this would require a knowledge of the entropies at the same internal energy, which is a very different thing.

The search must now be continued. Consider the polymerization again, but this time suppose it takes place in a bomb calorimeter. The walls of this are conducting and so the contents can be maintained at a constant temperature with the help of a suitable thermostat bath; but they are also rigid, and this means that (excluding from our argument for simplicity special forces like electrical and magnetic ones) no work can be done either on the system or by it. As a result of the reaction within the calorimeter a certain amount of heat will be evolved, and incidentally its measurement will be an easy matter. According to the First Law (see equation (2.1)), since no work has been done the amount of this heat will be equal to $-\Delta U$ for the substance in the calorimeter (the negative sign is necessary because it is heat evolved). Further, the substance within the calorimeter will have undergone a change of entropy, which can be called ΔS. Although the polymerization has been spontaneous nothing can be said about ΔS; it may be either positive or negative, since the system is not isolated.

We now turn to the Second Law. The process within the calorimeter, being spontaneous, is capable of being made to do work. Suppose an auxiliary apparatus is somehow introduced into the calorimeter by whose help the polymerization can be harnessed. It does not matter how this harnessing is done so long as it is carried out reversibly. The amount of heat which the calorimeter gives out to the thermostat bath is now no longer the value of $-\Delta U$ for the change. It has become instead an *absorption* of $T\Delta S$. This result may be unfamiliar, but it follows immediately from the definition of entropy. For if the heat absorbed by the calorimeter when the polymerization goes reversibly at temperature T is $Q_{\text{rev.}}$ then the entropy change has been defined as

$$\Delta S = \frac{Q_{\text{rev.}}}{T} \tag{7.1}$$

which gives

$$Q_{\text{rev.}} = T\Delta S. \tag{7.2}$$

We now invoke the First Law. In the (perfect) harnessing of the process the substance polymerized has given up a total amount of energy $-\Delta U$, measured by the first experiment. The bath, on the other hand, has given up a quantity of heat $T\Delta S$. Consequently the sum of these two $(-\Delta U + T\Delta S)$ by the First Law must represent the work done, and by Lewis's statement of the Second Law must be positive. Since the operation was reversible (or ΔS cannot be used in the way implied) this work will be a maximum and therefore for the transformation,

$$W_{\text{max.}} = -\Delta U + T\Delta S. \tag{7.3}$$

In any actual manipulation the process is harnessed less than perfectly, but whatever work is reclaimed it will obviously have the same sign as $W_{\text{max.}}$. Thus the very important conclusion can be drawn for a process happening spontaneously at constant temperature, that the sign of the quantity $(-\Delta U + T\Delta S)$ must be positive, for only then will it be true that work can be obtained from it.

Before this criterion is accepted as the one that is needed it can be put in a more convenient form. If the subscripts 1 and 2 denote the initial and final states of the system, and if a new function F is defined by the relation

$$F = U - TS \tag{7.4}$$

it follows that

$$\begin{aligned}-\varDelta U+T\varDelta S &= -(U_2-U_1)+T(S_2-S_1)\\ &= -(U_2-TS_2)+(U_1-TS_1)\\ &= -(F_2-F_1)\\ &= -\varDelta F.\end{aligned} \quad (7.5)$$

This criterion, that the quantity on the left hand side of this equation must be positive for an isothermal process to go by itself, can be put into the alternative form that $-\varDelta F$ must be positive, or F_2 must be less than F_1. The reader will have guessed that the function F is the free energy of the system, and the criterion arrived at is that for a process to be spontaneous under isothermal conditions, it must involve a decrease in free energy. Further, the decrease in free energy represents the maximum work that the process is capable of doing.

The Helmholtz and Gibbs Free Energies

The mention of free energy will probably make the physiologist feel more at home, for the notion is one that is of frequent occurrence in the discussion of his subject matter and it is certainly likely to be more familiar to him than entropy; but the very simplicity and familiarity of the idea is apt to hide some pitfalls. These will have to be discussed later; but before this is done there are some further matters which need attention.

When considering a biochemical or biophysical reaction in a cell or tissue it was noted that it occurs in a system which remains substantially unchanged in temperature while the reaction is in progress. This is not all; the system also remains substantially unchanged in pressure, usually that of the atmosphere. As an apparent exception to this latter statement it is easy to think of cases where for instance, turgor pressure increases, perhaps sharply; but nearly always this is due to material being taken into the cell, and the cell therefore is not the whole system. If the system boundary is drawn to embrace *all* the material substance concerned in the uptake process (i.e. to include part of the external medium) then the pressure across *this* boundary does remain constant. It is very exceptional indeed for processes occurring wholly within the cell to appreciably affect its turgor; hence the general statement stands. This constancy of pressure however adds a new element to the energy balance, for when a spontaneous process occurs in a cell, part of the free energy decrease (as defined above) is necessarily absorbed in doing work in the form of volume changes against this constant pressure. Still, the process goes on, it is supposed, in spite of this drain on its resources; Lewis's statement of the Second Law therefore implies that the process can still be harnessed to do work beyond this. If an argument is followed through

4

almost exactly on the lines of the preceding case (though it is a little more complex as an extra term is involved) the following conclusion is arrived at. If P is the external pressure on a system, and V is its volume, then if a new function is defined

$$G = U - TS + PV \qquad (7.6)$$

the condition for a process to go by itself in a system held at constant temperature and pressure is that its value of G should decrease; that is, ΔG for the process must be negative. The value of $-\Delta G$ measures what is often referred to as the maximum *useful* work, $-\Delta F$ measuring the maximum *total* work.

The two quantities F and G are in fact both spoken of as the free energy. The former is the Helmholtz free energy, and it is used as the criterion where no extraneous volume work is levied when the process takes place; it will be remembered in fact that in the argument leading up to this criterion it was supposed that the substance undergoing transformation was enclosed in a bomb calorimeter to ensure this. The latter function is the Gibbs free energy, and it is used as the criterion when the process has inevitably to do whatever work is necessary to promote volume changes against a constant environmental pressure. Since this condition corresponds most closely to the cellular situation, G is the function which is employed almost exclusively by biologists. In actual fact the value of ΔG for a metabolic or biophysical process is almost exactly the same as that of ΔF, except in the special cases when gases are involved. The difference between the two is of course $[PV]_2 - [PV]_1 = P\Delta V$ where the pressure remains constant, and only where gases are involved is ΔV worth taking into account. At least that is so unless we are experimenting at very high pressures, where a large value of P may redress the balance. Consequently the distinction between G and F, though important to others, need not concern the physiologist overmuch.

Direct Establishment of the Free Energy Concept

In introducing the idea of free energy it was based on the notion of entropy, and of course of energy and temperature. It is however possible to establish the idea more directly, without introducing entropy. This can be done instructively as follows.

Suppose, as usual, a closed system is considered; that is a definite fixed quantity of matter. Consider the transformation of this system between the initial and final states, which can be called (1) and (2) respectively. If the direction of spontaneous change is (1)→(2), this transition can be harnessed to do work; and if the transition is effected reversibly the work will be a maximum for the particular path chosen. However, other paths between the same terminal states are available (thus a salt may be split into its elements

7. ENTROPY AND FREE ENERGY

by electrolysis or by heating) and if one of these other paths is traversed reversibly the work will again be a maximum, but for the new path. However, these two maxima will not necessarily be equal. As we saw when discussing the First Law, while ΔU depends only on states (1) and (2), W depends on all the intermediate stages as well. The same therefore holds for W_{max}. But there is one case for which the matter can be put on a more definite basis. Suppose both the pathways involve no change of temperature. Then one pathway can be traversed reversibly and the work done stored (say by raising a weight); subsequently this stored work can be used to drive the system back along the other pathway. Unless the work involved is exactly equal in the two cases the balance (by the First Law) must be made up by a difference in the heat exchanged along the two pathways, since the system is restored to its initial state and therefore suffers no change in its own energy. All this is perfectly possible in the general case; but in the isothermal case under consideration the heat is taken in along one path and given out along the other *at the same temperature*. Thus the net result of the cycle would simply be to use heat from a single temperature and convert it to work[1]. This is contrary to the Second Law (see Kelvin's statement); consequently there is no balance, and the maximum work obtainable from an isothermal transformation must be quite definite and independent of the way the transformation is effected. This maximum amount of work can be defined as the free energy decrease associated with the change. In this argument it has been taken for granted that the pathways are isothermal throughout, and not just at their ends, so to speak. As a matter of fact this is a definite requirement, and it is easy to show that when a system changes between two states at the same temperature, but is allowed to wander in temperature during the process, the work it can do may have any value[2]. The same is true of course, when the terminal temperatures are unequal.

Free energy defined in this direct way is obviously just the same thing as the function defined in terms of entropy; the weakness of the direct method of definition is that it is less general in its scope. For instance, it does not indicate how the free energy change in a given reaction varies with temperature. This is an important consideration as will be seen later, and it can only be developed by introducing the idea of entropy.

AN EXAMPLE OF THE DIRECT USE OF THE FREE ENERGY

The argument of the previous section can be made a little more concrete, and at the same time an important result derived, by considering a specific example. Imagine a system consisting of a quantity of solution of concentration c at first confined under a piston of semi-permeable material

---

[1] For one direction of the cycle; vice-versa for the other.
[2] See Appendix II (p. 96).

above which is water (Fig. 6.1). If the area of the piston is A and the osmotic potential[1] of the solution is Π a force ΠA will have to be applied to the piston to keep the system in equilibrium. Imagine a very small quantity of water dn moles to be allowed to pass downward (by a very slight easing of the piston) into the lower compartment, the temperature being maintained constant by ensuring good thermal contact with the surroundings. If the passage is reversibly carried out the force on the piston remains at ΠA, and the latter moves up a distance dx, where Adx is the increase in volume of the solution. This increase in volume is not, in general, exactly equal to the decrease in volume of the water above; its value is $\bar{V}_w dn$, where $\bar{V}_w$ is the partial molar volume of water appropriate to the solution (Chapter 3). The decrease in volume of the water above the piston is correspondingly $V_w dn$, where V_w is the molar volume of water. The work done during the passage can now be calculated. The work delivered by the piston rod is $\Pi A dx = \Pi \bar{V}_w dn$ and this could all clearly be used to raise weights or drive a dynamo. In other words it is useful work, and no subtraction has to be made for work inevitably lost in expanding against the surroundings; consequently it represents $-\Delta G$. On the other hand, it must be noticed that it does not quite represent the total work done or $-\Delta F$. This latter has to take account of the small overall change in total volume, equal to $(\bar{V}_w - V_w)dn$. If $P_{atmos.}$ is the atmospheric pressure, the amount of work $P_{atmos.}(\bar{V}_w - V_w)dn$ would therefore have to be added to obtain $-\Delta F$. The point is instructive, but of no consequence, since for simplicity the assumption that the ambient pressure remains constant at $P_{atmos.}$ will soon be dropped.

The same free energy changes can however be computed using a quite different experimental programme, one involving passage into the vapour phase. To do this first imagine (the water and solution having been transferred to a more suitable apparatus) that the pressure on the water is reduced from atmospheric to its equilibrium vapour pressure p^0. It is important to realize that even this change involves a certain amount of work being done by or on the water and ideally this should be calculated. However, because water is so nearly incompressible it is permissible for these purposes to neglect it. In the next stage an amount of water dn is allowed to evaporate at constant pressure p^0 (the temperature is again kept constant). If v^0 is the volume of one mole of saturated vapour at p^0 the increase in volume is $(v^0 - V_w)dn$ and the total work done by the expansion is $p^0(v^0 - V_w)dn$; this represents $-\Delta F$. Since the pressure and temperature are constant (otherwise ΔG has no simple meaning) and since all the work is used up in expansion, there is no useful work left over and ΔG is zero.

---

[1] The value of Π depends very slightly on the ambient pressure, the effect being small owing to the virtual incompressibility of water. An exact treatment however would have to take this into account.

We now allow the vapour to expand isothermally from p^0 to a new pressure p, which is the value at which it will be in equilibrium with the solution. If the vapour is assumed to behave like a perfect gas, the work it does during this expansion is $RT \ln \frac{p^0}{p} \mathrm{d}n$ (see Chapter 5). This therefore represents $-\Delta F$. As a matter of fact it also represents $-\Delta G$, since the two differ only in the term PV which does not alter here (the expansion being isothermal and the gas being nearly perfect). However, it is necessary to be on guard; $-\Delta G$ in this stage has not the simple meaning of useful work, since although the temperature is constant, the pressure is not.

The next operation is to place the vapour at pressure p in actual contact with the solution at this pressure, the value p having been chosen so that the vapour and solution are in mutual equilibrium. Reversible condensation of the vapour into the solution is now accomplished by infinitesimally increasing its pressure. If v is the volume of a mole of vapour at pressure p and the prevailing temperature the work done *on* the vapour will be $p(v-\overline{V}_w)\mathrm{d}n$ since it condenses into a volume measured not by V_w, but by $\overline{V}_w$. With due regard for signs, therefore, for the last phase

$$-\Delta F = -p(v-\overline{V}_w)\mathrm{d}n.$$

As previously,

$$-\Delta G = \text{zero}.$$

Finally, the liquids have to be brought back to their original pressures, an operation in which as discussed above the work aspect can be neglected.

We now add up the changes in F and G as follows.

Stage	Operation	$-\Delta F$	$-\Delta G$
—	Alteration of pressures on liquids at start and finish	virtually zero	virtually zero
1	$\mathrm{d}n$ moles of water vaporized at pressure p^0	$p^0(v^0-V_w)\,\mathrm{d}n$	zero
2	Vapour expanded isothermally from p^0 to p	$RT \ln \frac{p^0}{p} \mathrm{d}n$	$RT \ln \frac{p^0}{p} \mathrm{d}n$
3	Vapour condensed at pressure p into solution	$-p(v-\overline{V}_w)\,\mathrm{d}n$	zero

Remembering that $p^0 v^0 = pv$, we have for the total changes

$$-\Delta F = RT \ln \frac{p^0}{p} \mathrm{d}n + (p\overline{V}_w - p^0 V_w)\,\mathrm{d}n$$

$$-\Delta G = RT \ln \frac{p^0}{p} \mathrm{d}n$$

and either of these expressions can be used to obtain a connection between the osmotic potential of the solution and its vapour pressure. Using the expression for ΔG as the simpler, and equating it to the alternative value found for the osmotic pathway we find

$$\Pi \bar{V}_w \, dn = RT \ln \frac{p^0}{p} \, dn$$

or
$$\Pi = \frac{RT}{\bar{V}_w} \ln \frac{p^0}{p}. \tag{7.7}$$

The reader may be puzzled to know why this same result does not come quite so easily if he attempts to use the values of ΔF calculated above. The reason lies in the various approximations[1] and omissions some of which have been indicated in the argument, but which in all ordinary cases are quite unimportant. It is purely a consequence of the method of introducing these approximations that it is found more straightforward at the end to use the expression for ΔG rather than the one for ΔF. In any case the extra term in ΔF is very small.

The result just obtained is a well-known one, and it will be derived later (Chapter 13) by a more sophisticated and more powerful method. The reason why it has been set out in such detail in this chapter is that it illustrates in a rather direct way how relations of this sort can be shown to depend on the fundamental laws of thermodynamics. Clearly, the argument employed to establish it could be recast in such a form as to show that if the relationship was not valid even in just one particular case, then it is at once possible, using this case, to design a piece of machinery to convert heat energy continuously into work without doing anything else to compensate for this conversion. Thus the validity of the relationship must be as wide as that of the Second Law. It is left as an exercise to the reader to show how a line of argument similar to the one employed to establish equation (7.7) could be developed to derive a relationship between the e.m.f. of a primary battery (a 'fuel cell') in which oxygen and hydrogen were consumed, and the equilibrium constant of the reaction

$$H_2 + \tfrac{1}{2} O_2 \rightleftharpoons H_2O.$$

As a guide of quite universal usefulness, Le Chatelier's Principle indicates how this reaction can be 'managed' reversibly; in the present case it suggests that the equilibrium can be made to lie as far over to the left as required by reducing the pressure, and as far over to the right as required by raising it. The further working out of the problem is left to the reader.

---

[1] For instance, not only liquid compressibility but also dependence of vapour pressure on the pressure of the liquid have been neglected.

Concluding Remarks on Free Energy

Before this rather discursive chapter on free energy is concluded some of the pitfalls and misunderstandings must be mentioned into which the physiologist with his specialist training is particularly liable to fall. The reader needs to be warned that even well-known biological textbooks often contain rather misleading statements on thermodynamics.

The first point to be noted about the free energy is that its change has a physical meaning only in situations where the temperature remains constant. This can be easily proved, for (dealing with F rather than G for simplicity) if a system changes from (U_1, T_1, S_1) to (U_2, T_2, S_2) then by definition

$$F_1 = U_1 - T_1 S_1$$
$$F_2 = U_2 - T_2 S_2$$

and
$$\Delta F = F_2 - F_1 = (U_2 - U_1) - (T_2 S_2 - T_1 S_1)$$
$$= \Delta U - \Delta(TS).$$

However, since classical thermodynamics does not enable us to know the *absolute* value of the entropy but only its *changes*, the value of

$$\Delta(TS) = T_2 S_2 - T_1 S_1$$

is undefined. If however $T_2 = T_1 = T$, then

$$\Delta(TS) = TS_2 - TS_1 = T(S_2 - S_1) = T\Delta S,$$

and ΔF becomes something quite definite. Moreover it has a simple significance, that of maximum work.

In the case of ΔG the above remarks still apply; but the additional point must be noted that while ΔG has a perfectly definite value even if P_2 is not equal to P_1, it ceases to have the simple meaning of useful work unless P_2 and P_1 are equal. Further, the pressure must remain constant throughout the process, and not merely be restored to its initial value at the end. These requirements must be carefully noted if ΔG is to be interpreted as useful work. Of course the insistence on constancy of temperature for both ΔF and ΔG does not mean that the free energy change of a reaction at one temperature cannot be compared with the corresponding free energy change at another. In fact this is frequently being done, as when the temperature-dependence of the e.m.f. of a battery is referred to, since the e.m.f. is a measure of the free energy change of the reaction involved.

A second point which must be noticed concerns a very frequent misunderstanding. The physiologist is used to speaking of the 'free water' and the 'bound water' in cellular systems; or of the 'free space' and 'osmotic volume' in uptake studies. In both these instances the 'free' component is part of the total, so it is natural if he comes to regard the free energy of the system similarly as being part of its total store, the other part being 'bound'. This way of looking at things is however inadvisable, and may lead to difficulties. For if the free energy is a part of the total energy then it must be subject, along with other forms of energy, to the Law of Conservation; if the free energy is allowed to run to waste without doing work, he may expect it to appear as heat, in an exactly equivalent amount. That this is sometimes not the case can be very easily shown, and in fact it is in general never the case. Suppose a balloon contains a perfect gas, and for the sake of simplicity suppose the skin is non-extensible and so stores or releases no elastic energy of its own. If this balloon is ruptured in a vacuum, the perfect gas escapes and fills the vacuum vessel; it does no external work, since it pushes nothing back. It gives up no energy and its temperature remains constant, this being a well known property of perfect gases. The gas has however suffered a sharp decrease in free energy. If free energy were just energy of a particular sort, then it might be thought legitimate to ask into what new kind of energy the wasted free energy had been converted, into 'heat' or into some form of potential energy? Obviously into neither of these, since no work has been done and the thermal energy of the gas is just what it was originally[1].

Suppose alternatively that the gas is allowed to expand to the same extent as before, again without change of temperature, but this time it is harnessed behind a piston. The same free energy decrease of the gas is now utilized for work; but where has the actual energy equivalent come from to do this work? This point has been discussed before. The gas gives up no energy, since its temperature remains unaltered. What happens is that the surroundings lose heat, and it is this heat which balances the work done. Thus from this alternative viewpoint, it is also apparent that the free energy decrease of the gas is not energy which enters into the 'balance sheet'. The work done is balanced by heat absorbed from the surroundings, and the free energy loss (or alternatively, as in Chapter 6, the entropy increase) appears rather as the price paid by the gas for the privilege of upgrading someone else's heat into work.

This is not the whole story, however. If the definition of free energy is recalled (using F rather than G for simplicity) then we may write

$$\Delta F = \Delta U - T\Delta S.$$

Free energy thus has two terms. In the case of the perfect gas discussed above, the first term (ΔU) is zero, and the whole of the free energy decrease is linked with the rise of entropy of the gas. Now entropy is not conserved; it can be created from nothing. That is why in this case all the free energy

[1] It is possible to say that the gas has done work 'on itself', and that this has then become converted into 'heat'. The device, however, seems a little artificial, if applied to the similar case of solute diffusion. It is not intended to imply however, that the division into 'free' energy and 'bound' or 'unavailable' energy is unworkable, but only that there are better ways of looking at things.

can be dissipated with no real form of energy to show for it. No actual energy belonging to the gas has in fact been either increased or diminished, but only its entropy. It is rarely the case that the value of ΔU is zero; perfect gases are the exception. When a change is dealt with whose free energy decrease involves a ΔU component as well as a $T\Delta S$ component then we have a free energy loss which does mean *real* energy loss (though the two are not exactly equal, but differ by $T\Delta S$). If in such a case the free energy is allowed to run to waste, the ΔU part of it must turn up as heat. But again the $T\Delta S$ part has no place in the energy balance sheet; energy has not got to be destroyed to create entropy, or vice versa. It is preoccupation with cases in which the ΔU term overshadows the $T\Delta S$ term that leads many physiologists to the conclusion that if the free energy is dissipated, an equivalent amount of heat must be produced. The oxidation of simple carbohydrates is a case in point, for here the first term (ΔU) far exceeds the second ($T\Delta S$). Consequently, unless the oxidation is harnessed there is a large quantity of heat liberated. The levelling-out of concentration differences is, on the other hand, a process whose ΔF depends much more on the second term, and correspondingly if the solute movements are allowed to occur unchecked there is little evidence of thermal effects.

The result of this discussion can perhaps be summarized as follows. The free energy of a system is best regarded as not being a part of its total energy, of which the other part is 'bound'. It cannot really be looked upon simply as energy at all, but represents rather a potentiality that the system has of doing work under certain circumscribed conditions, these conditions being that the system is supplied with all the heating or cooling it requires to keep its temperature unaltered. When the system actually does the maximum work at the expense of its free energy, part of the work is paid for out of its own resources (ΔU), and part is paid for out of heat energy supplied to it from its surroundings ($T\Delta S$). For the privilege of delivering this latter heat energy as work the system has something to pay, the price being a degradation in the form of an increase in its entropy. The two fractions comprising the work delivered may have varying proportions to each other. Sometimes the first is zero (perfect gases) and in rare cases the second may be zero, but in general both are present.

One final remark may be permitted on the consequences of there being two terms in the expression for the free energy (the extra term in ΔG makes no appreciable difference in condensed, that is non-gaseous systems). In the substantially isothermal systems with which cell physiologists have to deal, the state of equilibrium (as will be seen in the next chapter) is determined by the condition of minimum free energy. This implies two things: there is a tendency for the internal energy (U) to fall, and there is a tendency for the entropy (S) to rise; both tendencies lower F or G. The former is

4*

accomplished by suitable processes converting some of the energy to heat, which under the isothermal provisions is conducted away. (This would be ruled out under the conditions of isolation in a Dewar flask, and energy considerations would then have no effect on the equilibrium.) The latter is of course accomplished by processes which produce entropy, and these will be better understood after the chapter on statistical considerations. The point to be noticed here is that the two tendencies do not always work together, and the actual position of equilibrium will then be a compromise between them. A case in point is the Donnan equilibrium, which will be discussed here only sketchily. (Reference should be made to Chapter 12 and to Fig. 12.1 for the essential details of a Donnan system.)

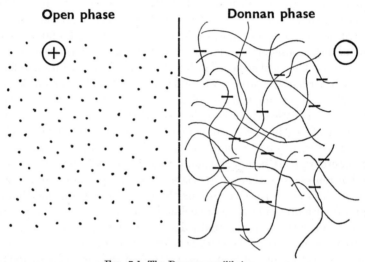

FIG. 7.1. The Donnan equilibrium.

Suppose the large immobile Donnan ions are negative. Then a natural potential difference will arise as indicated in Fig. 7.1 due to the tendency of the small accompanying positive ions to diffuse into the open phase. If a small quantity of a sodium salt is added to the system thus constituted, the question arises, how will the sodium ion distribute itself at final equilibrium? In the first place, there is an obvious tendency for any solute to achieve *equality of concentration* throughout, and this would actually happen in the case of a non-electrolyte like urea or sucrose. With the sodium ions there is a second tendency: they experience an electrical force urging them towards the Donnan phase, and if this was the only factor they would all end up there. What actually happens is that a compromise results, the sodium ions attaining a rather higher concentration in the Donnan phase than in the open one.

Thermodynamically the situation can be discussed in this way. Uniformity of concentration (as will appear from Chapter 9) is the condition of greatest entropy; or, if the expression may be permitted, it is what the entropy wants. However, accumulation of all the sodium ions in the Donnan phase is the condition of lowest potential energy; it is the conclusion the energy wants. In the case in which the system is held isothermally, a compromise is reached characterized by the condition of minimum free energy. The entropy has not risen to its highest value, nor has the energy fallen to its lowest, but the combination of the two represented by the functions G or F (according to circumstances) has attained its minimum.

Appendix I: Gibbs Free Energy and the Useful Work

It is a simple matter to prove that the decrease in the Gibbs free energy measures the maximum *useful* work obtainable under conditions of constant temperature and pressure. Consider a system which changes from (F_1, P, V_1) to (F_2, P, V_2). Then by definition

$$G_1 = F_1 + PV_1$$
$$G_2 = F_2 + PV_2.$$

Subtracting,

$$G_1 - G_2 = F_1 - F_2 - P(V_2 - V_1)$$

or

$$-\Delta G = -\Delta F - P\Delta V.$$

Now $-\Delta F$ is the maximum total work obtainable, and $P\Delta V$ is the work done by the system in expanding from V_1 to V_2 against the constant pressure P; hence the equation can be expressed verbally as

Gibbs free energy decrease = maximum work obtainable less the work done against external pressure.

In other words, the decrease in the Gibbs free energy represents the maximum surplus or useful work under the conditions of constant temperature and pressure. It is clear that the pressure must remain constant throughout the change, and not merely be restored to its initial value at the end. However, in contrast to temperature, it is not necessary for all parts of the system to possess the same pressure, since differences in pressure are not inconsistent with equilibrium. In such a case provided the individual pressures remain constant $-\Delta G$ for the whole system measures the useful work, but it must be remembered that the expression for G (that is, $U + PV - TS$) then contains several PV terms ($P^\alpha V^\alpha + P^\beta V^\beta + \ldots$), where the superscripts refer to the several parts or phases. For an example of this, refer to the end of Chapter 8.

APPENDIX II : DEFINITENESS OF THE WORK IN AN ISOTHERMAL PROCESS

When discussing the First Law it was seen that the terms W and Q for a given transformation (1)→(2) of a closed system were not defined unless the whole pathway of change was specified. One very important case in which the whole pathway *is* specified is in isothermal changes; and the argument used earlier to establish the free energy concept directly really amounts to this: that to say that a process is isothermal is sufficient to make W a definite quantity (perhaps it should be added that at least it is definite if the pathway is traversed reversibly, when of course W becomes W_{max}.).

That it is not enough merely to specify that the initial and final temperatures are the same can be seen by considering a Carnot cycle for a perfect gas (Fig. 7.2). This cycle consists of two isothermal stages AB, CD and two

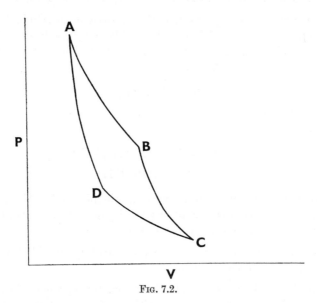

FIG. 7.2.

adiabatic ones BC, DA. Between the two states represented by D and C there are two pathways, one (DC) isothermal throughout, and the other (DABC) 'isothermal' only in the unorthodox sense that its end points have the same temperature. Clearly the work obtained along DABC exceeds that along DC by the area of the quadrilateral ABCD, and this can be made as large as necessary by proceeding far enough up the adiabatics. Thus unless it is stipulated that *isothermal* means *isothermal throughout* there is no definite maximum to the yield of work.

CHAPTER 8

Equilibrium and the Direction of Spontaneous Change

> 'What do Jagulars do?' asked Piglet, hoping that they wouldn't.
> 'They hide in the branches of trees, and drop on you as you go underneath,' said Pooh. 'Christopher Robin told me.'
> 'Perhaps we better hadn't go underneath, Pooh. In case he dropped and hurt himself.'
> 'They don't hurt themselves,' said Pooh. 'They're such very good droppers.'
>
> THE HOUSE AT POOH CORNER
> *A. A. Milne*

The idea of equilibrium, or thermodynamic equilibrium to be more precise, is one of the most important in the whole of thermodynamics. It has unavoidably been mentioned already, but in this chapter it will be discussed more fully and systematically.

The notion of equilibrium is one which is fairly easy to appreciate intuitively. It denotes a state of rest, of balance, of absence of change, and it arises in the first place in connection with systems most aptly described as mechanical. An object falls over, rolls about for a time and finally comes to rest; we say that it is then in equilibrium. Or a commodity may be thrown into the pan of some spring scales; the pan oscillates up and down for a few seconds and then settles down to its position of equilibrium. Sometimes the idea is a little less clear cut: an acrobat may balance on a ball and remain virtually stationary, but probably he would hardly describe his state as one of rest, though he might call it equilibrium.

As it is defined in thermodynamics the idea of equilibrium may be said to involve several requirements. In the first place, for a system to be in equilibrium it must possess properties which are not changing with time. By 'properties', as always in thermodynamics, are meant characteristics which are observable and measureable on the macroscopic scale, such things as density, temperature, turgor, state of magnetization and so on; we do not include features of the sub-microscopic world of molecular Brownian movement such as the position of individual molecules. In the second place, the system must be at rest, in the sense that its condition is not being maintained by the continuous expenditure of energy or effort in any form. To make this second requirement clear, consider the glowing filament of an electric lamp. It is in a condition which is not observably altering from moment to moment, but its state is certainly not one of equilibrium; the filament is receiving and giving out large amounts of energy all the time, just as a leaky tank into which a tap is running may be receiving and giving out large amounts of liquid substance. Consequently, while the filament satisfies the first condition that is laid down it does not fulfil the second, and it is referred to as being in a *steady* or *stationary state* rather than in equilibrium.

Another way of expressing the second requirement is to say that for a system to be in thermodynamic equilibrium there must be no macroscopic, observable processes taking place in it. Brownian movement, even if it could be considered as macroscopic, has no 'direction' associated with it. Consequently it is not a process in the sense in which that word is used, and its observance is not incompatible with equilibrium. In fact, the postulate of chaotic molecular movements (the epithet 'Brownian' can be extended to these, though originally it was applied to particles large enough to be visible in the microscope) forms an essential element of the study of equilibria in the statistical development of the subject. As a result it has come to be realized that equilibrium in thermodynamics is essentially a dynamic affair; the molecules are not stationary, like the elements of a mechanical system at rest, but are in the condition in which their movements balance out. As often as two molecules react chemically to form a third, so (on the average) does one of the third type split up to form the two. And the same thing is true of transport and other processes when equilibrium prevails. Unfortunately the phrase 'dynamic equilibrium' is often used to describe not this aspect of things but rather the condition spoken of as a 'steady state'. This is to give it an essentially different meaning, and the reader has to be on his guard as to what is being intended. A steady state is of course dynamic, but on the macroscopic level. On the other hand, a state of equilibrium, in the thermodynamic sense, is also a dynamic affair, but on the microscopic level.

Types of Equilibrium

It is convenient to distinguish several different types of equilibrium, and these can best be illustrated with reference to mechanical analogies. A ball resting in a hollow (Fig. 8.1(a)) represents *stable* equilibrium; a slight displacement calls into play forces tending to restore the original configuration. The same ball resting on the summit of a hillock (Fig. 8.1(b)) is in *unstable* equilibrium for contrasting reasons; while if it is on a horizontal plane (Fig. 8.1(c)) it is said to be in *neutral* equilibrium.

Two further useful categories are *metastable* equilibrium (Fig. 8.1(d)) and *false* equilibrium (Fig. 8.1(e)), the qualifications being that in the first, equilibrium is stable only provided the displacements are not too big, and in the second that the system is really moving, but too slowly to be perceptible. Of these different types of equilibrium (a) is of course quite common in

8. EQUILIBRIUM AND THE DIRECTION OF SPONTANEOUS CHANGE 99

thermodynamic systems, while (b) represents an impossible state of affairs on account of Brownian movement. Neutral equilibrium (c) might be instanced by the case of a liquid and its vapour filling a cylinder closed by a piston, the pressure of the environment being equal to the vapour pressure. Under these conditions if the piston is pulled out slowly, more liquid evaporates

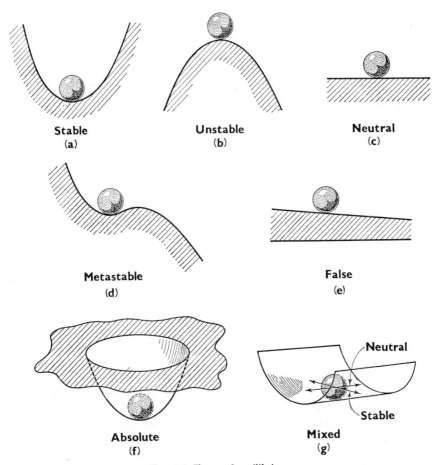

FIG. 8.1. Types of equilibrium.

to maintain the vapour pressure, and on release the piston merely 'stays put'; the temperature is assumed to remain constant. An example of metastable equilibrium is perhaps provided by a supercooled liquid, or a supersaturated solution; and of false equilibrium by a solid like glass, subject to extremely slow crystallization.

Mechanical analogies can however be carried further. It is understood in Fig. 8.1(a)-(e) that the ball is subject to displacement in the plane of the paper. Suppose, however, that it has another degree of freedom and can move perpendicular to the paper (Fig. 8.1(f) and (g)); in this direction its equilibrium may have quite different characteristics. A ball resting at the base of a hemispherical depression (Fig. 8.1(f)) is in stable equilibrium no matter in what direction it is moved, and its equilibrium may be described as *absolute*. However, if the depression is shaped like a valley (Fig. 8.1(g)) the case may be different. If the floor of the valley is level the equilibrium will be stable for displacements across the valley and neutral for displacements along it; such an equilibrium may be described as *mixed*.

As an aside in our discussion, if the floor of the valley is sloping the ball may be visibly moving down it. Whether this movement appreciably affects the position of the ball at the lowest point of the valley cross-section depends on circumstances. One of these is the speed of its movement down the valley relative to the speed at which the ball moves to the lowest point when displaced transversely. Another is whether the valley floor is straight or curved, for if it is curved then a centrifugal force will tend to throw the ball up the valley sides as it rolls. These two considerations can be summed up like this: the movement towards one equilibrium (rolling down the valley) will affect the position of another equilibrium (settlement at the lowest point of the cross section) to an extent dependent on the relative speeds of the two movements, and to what might loosely be called the linkage between them. If the movement down the valley is very slow, then it hardly matters how the path twists or turns; the ball will scarcely move from the lowest point of the cross-section, especially if the valley is steep-sided and the lateral recovery is therefore quick. Alternatively, even if the valley inclines sharply it will make no difference provided it is quite straight. In this case it may be said that there is no linkage between the two directions.

Thermodynamic systems have of course many more degrees of freedom than the simple analogy just discussed, but the principles which apply are the same. Two instances of equilibria that are attained very rapidly are the random Maxwellian pattern of molecular movements, and ionic dissociation. Two reached much more slowly are the chemical weathering of rock and the diffusion of a solute in water. Consequently, to take the last, lack of diffusional equilibrium will hardly influence to a significant extent the statistical pattern of thermal molecular velocities, or the pH of the solution in which the diffusion is taking place. The analogy also shows that even if two processes move at comparable rates it does not follow that there will necessarily be an interaction between them, though one must always be on the look out for it. This whole subject is of course allied to the physiological concept of 'active transport', for the latter is nothing else but the influence of non-equilibrium in one process on the position of rest of another. The rolling of the ball down the valley, to translate the analogy into the terms of an example well known to physiologists, might stand for the degradation of carbohydrates in the process of salt respiration. As the above discussion has shown, using even the very limited versatility of the mechanical model, such a degradation might drive another process away from its equilibrium state, in particular promoting an accumulation of ions against a concentration gradient, a 'rolling of the ball up the valley sides'. It all depends on whether the two processes are so related that there exists a linkage or interaction between them; and this is a matter to be explored not by thermodynamics but by an investigation of mechanism, usually on the molecular level. Further discussion of active transport must be deferred till a later chapter.

Types of Process Concerned with Equilibrium

It is possible to classify types of equilibrium in a different way, namely with respect to the kinds of process involved. Thus it is possible to speak of diffusional equilibrium, which normally obtains when the concentration is everywhere the same; or of phase, chemical, electrical, thermal, osmotic, mechanical or radiative equilibrium. All of these imply conditions of their own analogous to that expressed for diffusional equilibrium, but each with its own quite distinctive character. Thus thermal equilibrium requires uniformity of temperature, and mechanical equilibrium, balancing of forces. It is one of the functions of thermodynamics to help us to make comparable statements about other types of equilibrium; for instance, what is it that has to be 'uniform' or 'balanced' in chemical or osmotic equilbrium? This point will be taken up later; but while the subject is being considered there is a matter of interest, relating to the types of process listed above, which is worth mentioning in passing. Most of those listed are vector processes[1]; that is, they have associated with them a direction in space. Chemical processes particularly (among those mentioned) have a scalar quality, though some of the others may on occasion fit into this category too; and conversely it is not impossible for chemical reactions to be directed in space. However, the matter to be noticed is this. In connection with active transport (by definition a vector process) only a vector force[2] can provide the immediate driving power. Unless it is possible to invoke a vectorial chemical reaction (such as might conceivably take place on the cell membrane across which transport is occurring) a vectorial process such as diffusion of a carrier substance must be assumed. If this ultimately has to be sustained by a scalar metabolic reaction

---

[1] For instance diffusion, heat conduction and the flow of electricity are vector processes. Fusion, solution and chemical reaction are scalar processes.

[2] 'Force' is used here in a wide sense (see footnote 1, p. 235).

then the vector character must be provided for by assuming that this reaction occurs unequally at two different sites, one on each side, say, of the membrane. These considerations are usually not violated by those who speculate on the mechanism of active transport, but it is worth making explicit what is implicit in most speculations, that vector processes can *directly* interact in the sense previously discussed only with other vector processes, and likewise scalar only with scalar.

Absence of Physical Barriers

When speaking of a system as being in thermodynamic equilibrium it is taken for granted that, so far as the type of process under consideration is concerned, no physical barrier is present to its actual occurrence. Thus a system of two solutions of unequal strength not in contact could hardly be described as being in osmotic equilibrium, because a physical barrier, perhaps the sides of glass bottles, is interposed between them. It is instructive to inquire just what is meant by the absence of a physical barrier, or how the decision can be made that one is absent. As is so often the case in this subject the answer can be given on two levels: on the macroscopic level (obviously the right one for classical thermodynamics), and on the microscopic or molecular one (appropriate to the statistical development). To take the second

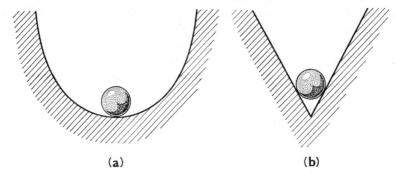

(a)   (b)

Fig. 8.2. Representation of thermodynamic equilibrium.

level first, the criterion that no physical barrier to osmotic equilibrium exists is that the osmotic process (passage of solvent molecules) is itself actually taking place but at rates equal and opposite for the two directions. Clearly this rules out the case of the separately bottled solutions. Since classical thermodynamics knows nothing of molecular theory and in fact deals only with properties and processes which can be plainly observed, such a criterion is foreign to it and another is required. This is provided by the requirement that the smallest possible force on the separating membrane (a membrane is

8. EQUILIBRIUM AND THE DIRECTION OF SPONTANEOUS CHANGE   103

implied in osmosis), or alternatively [1] the smallest possible change of pressure on one of the solutions, will cause an observable passage of solvent. This criterion will only be valid if, to refer again to a mechanical analogy (Fig. 8.2(a)), the base of the hollow is flat and horizontal, for only then will the smallest horizontal force produce a measureable displacement. This brings us to an important point. Thermodynamic equilibria are in fact represented by such a model as Fig. 8.2(a) and not by one such as Fig. 8.2(b). This can be accepted as a fundamental fact of experience, and its importance will at once be apparent. Were it not the case, a system displaced infinitesimally by an infinitesimal change in the reactions of its surroundings (for example, in their pressure or temperature) could not be postulated.

ANALYTICAL CONDITIONS FOR EQUILIBRIUM

The conditions for equilibrium can now be set down in terms of the thermodynamic functions previously discussed. In deriving these conditions we compare the values at equilibrium of those properties of the system which are believed to be relevant with the values they would have if the system were a little different. In other words, we imagine the system to be 'displaced' from its equilibrium state and examine the effect of this on its entropy, or free energy or perhaps some other function. For instance, if the system contains a set of chemical substances such as starch, water, phosphoric acid and glucose 1-phosphate which interact according to the equation

$$\text{starch} + \text{phosphoric acid} \rightleftharpoons \text{glucose 1-phosphate aq.}$$

then we imagine it to be displaced in such a way that more of the reactants on one side of the equation are present and less of those on the other. Or if the system contains a membrane and diffusing substances, then we imagine some of the latter to be transferred one way or the other across the membrane. As a final example, if the system contains both a liquid and its vapour phase then we imagine more of the vapour to be present and less of the liquid, or vice versa.

At this point it is instructive to pause and consider how such imaginary displacements could in fact be carried out. When any change takes place in a closed system, in general there will always be alterations in volume and temperature either of the whole system or of parts of it[2]. This is sometimes very obvious, as in such a highly exothermic chemical reaction as the combustion of hydrogen; but it is also the case in such unspectacular changes

---

[1] Alternatives are necessary because the details of the container have not been specified. The first applies where the membrane is piston-like.

[2] Or perhaps it is better to say, tendencies to alteration, since by suitable arrangements T and V can be held constant.

as the disappearance of concentration gradients or the phosphorylation of starch. What it implies however (on the grounds of Le Chatelier's Principle) is that a means is always to hand of displacing the completed reaction either backwards or forwards by the simple expedient of manipulating the pressure or temperature. Where the process is a scalar one (e.g. a chemical reaction) the manipulation is made on the whole system; where it is a vector one (e.g. diffusion) it has to be applied vectorially, that is, one part of the system is manipulated differently from another, and the element of direction in space is introduced. It is left as an exercise to the reader to apply these principles to such reactions as those mentioned: the phosphorylation of starch, of the diffusion of a solute in water (in the latter case the localized application of pressure requires the introduction of an auxiliary membrane, differentially permeable).

These considerations raise another point, however. When Le Chatelier's Principle is taken advantage of and the temperature or pressure of a system is changed in order to 'displace' it, the system simply moves to a new position of equilibrium. What we wish to do is to remove it from equilibrium altogether. This can be done if, having displaced it, auxiliary constraints are applied to 'freeze' it in its new configuration while the initial conditions, of temperature and pressure say, are restored. These auxiliary constraints may take the form of thin partitions which are slipped in to prevent displaced solute from diffusing back when a local increment of pressure is relieved; or they may be catalytic poisons in insignificant quantities which will hinder the return movement of a chemical reaction. This may seem rather an artificial procedure; but its justification lies in the fact that it can often be done in real life, and this encourages the belief that it is not nonsensical to invoke it in general.

The system is now genuinely out of equilibrium, and its new entropy and free energy can be calculated and compared with the original values. But first, what is meant by the 'initial conditions' to which the system has been restored? If the system was displaced by changing its temperature then the obvious answer would seem to be, to return the system to its initial temperature; but this will not in general be to its original pressure or volume or internal energy. It can perhaps be arranged for two or even more of these parameters to come back to their original values, but it cannot be arranged for them all to do so; and this leaves a choice as to which in fact shall be restored. Although the temperature may have been manipulated to effect the displacement, temperature has no prior claim to be the parameter to be restored to its original value, since the same displacement could have been achieved in other ways. Thus it comes about that there are quite a number of different conditions under which the comparison between the system in equilibrium and the displaced system can be made, and these provide the alternative criteria for the equilibrium state.

Mathematical Expression

The mathematical formulation of these considerations is obtained as follows. First, we suppose that the displacements made are extremely small, in fact infinitesimal. This is done because it simplifies the mathematics for reasons which have already been discussed. Second, we notice that the overall process, in which the displacement is effected, auxiliary constraints are applied, and the initial conditions are restored, is a reversible one, since the system moves successively from one position of equilibrium to another only infinitesimally removed from it[1]. Considering next the method of effecting the displacement, it will be obvious at once that in general heat (dQ) and work (dW) will have passed between the system and its surroundings. Individually, dQ and dW will depend on the method of carrying out the displacement; but in all cases the First Law will apply and the equation can be written (with the usual convention as to signs, see Chapter 2)

$$dU = dQ + dW \qquad (8.1)$$

where dU is the increase in internal energy of the system. We now introduce the condition that the infinitesimal change is a reversible one. If dS is the increase in entropy of the system, then by definition

$$dS = \frac{dQ}{T}, \qquad (8.2)$$

and this enables the substitution to be made

$$dQ = TdS. \qquad (8.3)$$

Further, if a case is being dealt with in which the surroundings do not exert on the system any of the less usual forces such as electrical or magnetic ones, the work done by the surroundings will be simply that associated with volume changes, namely $-PdV$. The negative sign is of course required by the fact that positive work on the system implies a diminution in its volume. Introducing these changes gives for a simple system displaced infinitesimally from equilibrium

$$dU = TdS - PdV. \qquad (8.4)$$

This is the equation which forms the starting point for discussion of the equilibrium of a closed system.

Before developing the implications of this equation we will pause for a moment for a closer look at it. It is worth while re-emphasizing that the equation applies only to a *closed* system (see Chapter 2), and equations applicable to open systems will have to be developed later. Then, unless further work terms are added to the right hand side it applies only where the

---

[1] This assumes that thermodynamic equilibrium is like Fig. 8.2(a) and not like Fig. 8.2(b).

surroundings exert no forces on the system other than those associated with pressure. Thus electrical forces are excluded; but this does not mean that they cannot be present and indeed important *within* the system, as they often are in cells. So long as the boundaries are so drawn that the electrical forces do not operate across them they will have no contribution to make to dW and can be forgotten. Of course it is easily possible to allow for external electrical and other forces if required.

It is not easy to give a satisfactory mechanical analogy of equation (8.4), but the following may help. Imagine a small heavy body of mass m and weight mg acted on by a number of forces F_1, F_2 and F_3 (Fig. 8.3), which

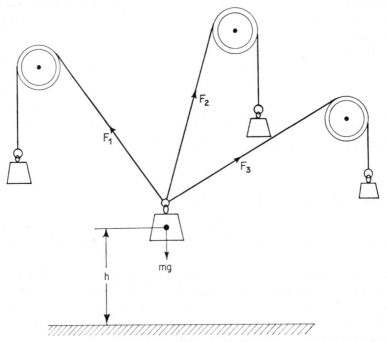

FIG. 8.3. A mechanical analogy.

might be applied through strings running over pulleys and carrying weights. Suppose the body is in equilibrium under the action of these forces. Then it can readily be appreciated that if an attempt is made to displace it by hand the *smallest measurable force* (barring friction) will cause displacement. This is in line with what was said earlier about thermodynamic equilibrium.[1] Under these conditions (i.e. presupposing equilibrium) the algebraic total of the work done by all the forces F_1, F_2, F_3 and mg in a minute displacement is zero, a result known as the Principle of Virtual Work which can be

[1] Compare Fig. 8.2 and corresponding text.

8. EQUILIBRIUM AND THE DIRECTION OF SPONTANEOUS CHANGE

appreciated intuitively. In symbols this can be written

$$mg(-dh) + F_1 dr_1 + F_2 dr_2 + F_3 dr_3 = 0 \qquad (8.5)$$

where $-dh$, dr_1, dr_2, dr_3 are the displacements measured in the directions of the corresponding forces. Now $mgdh$ is the increase in the potential energy of the body; calling this dU gives

$$dU = F_1 dr_1 + F_2 dr_2 + F_3 dr_3 \qquad (8.6)$$

and this equation is a fair analogy to equation (8.4). Its derivation illustrates what is meant by 'displacement', and also the fact that the choice of the direction (giving this word a rather wide meaning) of the displacement is immaterial. It suggests that temperature and pressure [1] are rather like the tensions holding the body at rest, and that the changes in entropy (dS) and volume ($-dV$) are the 'displacements' measured in the 'directions' of the temperature and pressure constraints. This is in fact rather how Brønsted[2] views matters. The equation of virtual work is used by engineers not so much to determine whether a system is in equilibrium, as given equilibrium, to derive relationships between the forces. This is in fact just how the thermodynamic relationship is used; but in the latter context it is also often interesting to know in what direction equilibrium lies, and equation (8.4) can be developed to give this information.

Alternative Statements of the Conditions of Equilibrium

The following is really only a formal mathematical development of equation (8.4). The two free energy functions, F the Helmholtz free energy and G the Gibbs free energy were defined previously (Chapter 7) as follows:

$$F = U - TS \qquad (8.7)$$

$$G = U + PV - TS. \qquad (8.8)$$

A further useful function is the enthalpy H. This is often rather misleadingly called the 'heat content', and in spite of its exotic sound the term enthalpy[3] is preferable. It is defined as

$$H = U + PV \qquad (8.9)$$

so that we can write the further equation

$$G = H - TS \qquad (8.10)$$

which means that G has the same relation to H as F has to U.

---

[1] Sometimes the thermodynamic equation has to include other terms to allow for the doing of electrical, magnetic or other forms of work. These can be accommodated by extra 'pulleys', and in fact one extra has already been allowed (cf. (8.4) with (8.6)).

[2] For full reference see footnote three p. 13.

[3] It is easy to show that the value of $-\Delta H$ for a change represents the heat evolved in a constant pressure calorimeter, as $-\Delta U$ does in a constant volume (bomb) calorimeter.

If the first three equations above are differentiated we obtain
$$dF = dU - TdS - SdT \tag{8.11}$$
$$dG = dU + PdV + VdP - TdS - SdT \tag{8.12}$$
$$dH = dU + PdV + VdP. \tag{8.13}$$

Adding these three purely mathematical identities to the thermodynamic relation (8.4) gives in turn
$$dF = -PdV - SdT \tag{8.14}$$
$$dG = VdP - SdT \tag{8.15}$$
$$dH = TdS + VdP \tag{8.16}$$

and these three equations stand on exactly the same footing as the original one (equation (8.4)), only involving different variables of state. Their value lies in the fact that they enable us to see what happens to the system when, after displacement from equilibrium by taking advantage of Le Chatelier's Principle, it is restored (before comparison is made with its original properties) to differently chosen 'initial conditions'. This point, it will be remembered, was raised earlier in the chapter, and we proceed to it at once.

Suppose for instance that the displaced system is restored to its original internal energy and volume. Bearing in mind that all external forces except pressure are excluded this means, in effect, working with an isolated system because it is implied that no work is done on it, and with unchanged internal energy there can have been no intake of heat, by the First Law. Now analytically, constant internal energy and volume means that both dU and dV are zero. The only one of the four equations (8.4, 8.14, 8.15 and 8.16) in which these two differentials occur together is the first. Giving them both the particular value zero it follows at once that dS must also be zero. In other words, compared with every other slightly displaced state of the system for which the energy and volume are the same the equilibrium condition has stationary entropy. This particular property then is either a maximum or a minimum, and it is not difficult to settle by further considerations that it is a maximum; in fact this conclusion was virtually arrived at in Chapter 7.

In an exactly analogous way other criteria can be obtained. Suppose the system is compared with its state when displaced a little but brought back to the original temperature and volume. Here dT and dV are each made equal to zero. The relevant equation is (8.14), and it shows at once that dF is zero, or F stationary. It is easy to see that F is in fact a minimum when a comparison is made in this way. If on the other hand the temperature and pressure are kept constant, equation (8.15) shows that it is G which is stationary, in fact minimal. The fourth equation (8.16) while theoretically interesting, does not lead to such a generally useful criterion, so further consideration of it will be omitted.

Applications to Living Systems

Now that three alternative criteria of equilibrium have been established we are in a position to see how they can be applied to physiological situations. As a matter of fact they are very rarely used directly as criteria for equilibrium; they have rather a different function. Recalling the definition of stable equilibrium given earlier it will be remembered that the term implies that thermodynamic systems always move *towards* equilibrium. Coupled with the appropriate criterion that equilibrium is a condition of maximum entropy, or minimum free energy, this implies that spontaneous changes are necessarily in the direction of increasing entropy or decreasing free energy, according to circumstances; and it is in this recognition that the real use of the criteria lies. They indicate in which direction biological systems will move spontaneously.

In deciding which criterion to use convenience is naturally the arbiter; and in this connection it is to be noted that when changes take place in biological systems the action of the environment normally maintains the temperature and pressure substantially constant. Under these conditions therefore the Gibbs free energy is appropriate; and thermodynamics shows that it must necessarily decrease in all spontaneous happenings. It is important to note however, that under these typical conditions of constant temperature and pressure no obvious statements can be made about the entropy or Helmholtz free energy of the system; they may go either up or down. Only if the other appropriate parameters are held constant does the behaviour of these become unequivocal.

The way in which the free energy criterion is used may be illustrated by a few examples. The first concerns the fixation of atmospheric nitrogen by bacteria; is it a reaction which requires energy (or strictly, work) to promote it, or can it on the other hand be looked on as a process capable of energizing others? This problem is handled as follows. We first write down the overall equation describing the process (this will in fact be done in detail in a later chapter for a similar example, but here the concern is only with principles). Having done this we obtain from published tables figures for the free energies of the various reactants and products in the states (e.g. gaseous or dissolved, at specified pressures or concentrations) in which they actually appear in the equations. Summation then gives the total (Gibbs) free energy relevant to the two sides of the equation, and we discover whether there has been an increase or a decrease as a result of the fixation. If there has been an increase then the fixation is not self-supporting energetically (i.e. it is 'endergonic') and another process must be sought that releases free energy (i.e. involves a free energy decrease, or is 'exergonic') to drive it. If on the other hand nitrogen fixation proves to be a spontaneous or exergonic reaction then it could conceivably provide the driving force itself for another process, such as protein formation, in which there is an increase of free energy. The only requirement which thermodynamics lays down is that the extent of the protein formation (say) must not be so great that on balance, the two linked processes involve an overall increase in the free energy. Provided there is a net decrease, thermodynamics raises no objections (though of course biochemistry may do so).

The second example concerns redox potentials. In the case of many oxidation–reduction reactions it is possible by balancing their 'urge' or 'affinity' (Chapter 11) against an electrical potential difference to make a direct measurement of the free energy change involved. This follows from the facts firstly, that the free energy change represents the maximum useful work that can be obtained from the reaction (T and P being constant); secondly, that the potential difference can be measured while drawing such an infinitesimal current that the reaction proceeds substantially reversibly; and thirdly, that the potential difference is itself the work associated with the passage of a unit quantity of electricity. As a consequence if a table [1] is drawn up showing the redox potentials of all the systems present in the cell it can be seen at once whether one particular system is able to reduce another; it is only necessary to ensure that its redox potential is greater. This may seem obvious, but perhaps what is not quite so obvious is that it is an application of the free energy principle. Of course, in comparing the redox potentials it must be ensured that they apply to the chemical systems as they actually exist in the cell; that is loosely, with the reactants at their actual concentrations.

The third example concerns a physical process. It has sometimes been conjectured that active cellular mechanisms assist in the raising of water to the tops of tall trees. Has thermodynamics anything to say on this point? The overall process in this case is the conversion of water from the form of a dilute solution in the soil to that of unsaturated vapour, and it is a simple matter to calculate the free energy change involved. This is a case incidentally in which the Gibbs free energy differs widely from the Helmholz free energy, since a large alteration in volume is concerned. The decrease in the Gibbs free energy represents the *maximum* useful work that can be done; consequently it must exceed the actual work done. This latter comprises the work done against gravity in raising the water, and the work done against the frictional forces of viscosity during its passage through the root cortex and vessels. Some means would have to be found for estimating this frictional work. When finally the comparison is made, if the free energy decrease does not exceed the total work done then clearly exergonic processes must be sought besides that of the vaporization to redress the balance; and it would be fashionable to postulate some component of the respiration.

[1] See, for instance, Dixon, M. (1949). 'Multi-Enzyme Systems', p. 73. Cambridge University Press, Cambridge.

8. EQUILIBRIUM AND THE DIRECTION OF SPONTANEOUS CHANGE

This last example has been chosen because it can be used to illustrate several points. The Gibbs free energy, it will be recalled, measures the maximum useful work which can be obtained in a process when the temperature and pressure remain constant. Is that the case here? The temperature will be considered in a moment, but meanwhile the transformation of the water may be looked on in the following light. A quantity v_l of soil water is eliminated, as it were, from the soil into the root and in this process the pressure P of its surroundings (substantially atmospheric) does an amount of work Pv_l upon it. This same quantity of water is then lifted and evaporated from the leaf, assuming a volume v_g at a partial pressure p. In this evaporation it has to do the work pv_g in 'pushing back' the water vapour already present in the atmosphere. It should be particularly noted that the pressure involved here (p) is the partial pressure of the vapour, not the full pressure of the atmosphere[1]. The net result is that the amount of work $pv_g - Pv_l$ has been unavoidably lost owing to volume changes and it is this loss that the use of G rather than F takes into account. It must be remarked that 'constancy of pressure' does not imply that if $-\Delta G$ is to be given its meaning of maximum useful work then P and p must be equal; all it requires is that P must be constant at the roots during the uptake of water, and p constant at the leaves during its vaporization and elimination. Further, these values of P and p must be used in calculating the Gibbs free energies before and after the movement respectively. This answers the first question.

In the second place, what about the temperature? It is common knowledge that the temperatures of the leaves and roots of trees are often widely different. Now the requirement that the temperature must remain constant is not met (as in the foregoing case of pressure) if the leaf and root environments remain individually steady; the temperature must be constant throughout the whole system as well, or else the free energy change means nothing, as was seen in Chapter 7. The answer therefore is that the use of the free energy criterion *is quite invalid* unless there is complete uniformity of temperature, a requirement that can be very easily violated in the case of extended systems like trees.

The last points to be remarked upon are somewhat different. Assuming uniformity of temperature a decrease in the free energy of the water from soil to atmospheric vapour indicates that evaporation can take place spontaneously; but the water has not merely to be evaporated, it has to be lifted as well. This means that the overall value of $-\Delta G$ must not merely be positive (as in the other examples). It must exceed the gravitational work done, and this is a new feature which arises because gravitational potential energy is not

---

[1] Macroscopically, the opposition the evaporating water meets is from the vapour already there. It is this which must be 'pushed back'; the oxygen and nitrogen remain where they were during this process.

usually reckoned as a part of the internal energy U. Hence it has to be taken account of separately, and again it puts in an appearance with large systems, like trees.

Finally, it occasionally happens that when an overall $-\Delta G$ has been calculated and found to allow a process to be spontaneous, the possibility still exists of making an estimate from the measured rate of the process of the energy requirement of some of the irreversible ('frictional', etc.) losses. Now if these can be estimated and deducted from the calculated positive free energy decrease then clearly there must still be a little of the latter left over to take account of further incalculable losses. This extension is clearly going beyond a strictly thermodynamic procedure, and it so happens that it can be illustrated by the case under discussion. Thus with a knowledge of the rate of flow of the transpiration stream, and of the pressure drop in the vessels due to viscosity and friction the dynamic work requirement for maintaining the flow of water in the vessels can be estimated. Therefore, the final criterion which is applied to ascertain whether or not respiratory energy must be invoked to account for transpiration at the measured rate, takes the somewhat unusual form

$$-\Delta G > \text{gravitational work} + \text{work spent in overcoming viscosity and friction.}$$

Naturally enough the possibility of calculating the irreversible consumption of free energy in this way and allowing for it is rather more likely with gross phenomena like transpiration than with fine-scale ones like metabolic reactions. This means that in the latter case the free energy criterion can only tell us whether a process is possible, and not whether it is possible *at a certain rate*. It is this additional information which can be obtained in the transpiration case just discussed, and this is because it is possible to estimate the 'frictional' losses as a function of rate.

NOTE ON MECHANICAL ANALOGIES

Thermodynamic situations can never adequately be represented by mechanical analogies. Commonly it is the aspects concerned with entropy that suffer most. Thus in equation (8.4) the terms on the right are qualitatively quite different; in the mechanical equation (8.6) they are all but identical (i.e. all are the product of a force and a distance). Further, it is clearly the 'temperature × entropy' term which is least at home in the mechanical illustration.

This means that one must not press such analogies too far or they become misleading. Those given later (pp. 253–4, 256–7) need especially to be viewed in the light of these remarks.

CHAPTER 9

The Statistical Interpretation of Equilibrium and Entropy

> Now it happened that Kanga had felt rather motherly that morning, and Wanting to Count Things—like Roo's vests, and how many pieces of soap there were left.
>
> THE HOUSE AT POOH CORNER
>
> A. A. Milne

However great an effort is made to make the more abstruse ideas of thermodynamics comprehensible by arguments confined to the world of large-scale events it is only when we turn from these to consider the molecular nature of things that any real success is achieved. In this chapter therefore some very simple statistical illustrations will be used to try to clarify the molecular foundations of the subject.

A STATISTICAL MODEL

Imagine a hard flat table on which there are a number of identical counters; suppose their number is one hundred. The counters are uniform and symmetrical, except that one side is black and the other white. Further, imagine them to be endowed with the property of jumping in a spontaneous and random fashion, so that sometimes they land with their white surface uppermost and sometimes with their black; after a short interval they jump again. Naturally, once they start jumping each individual counter will in the aggregate spend half of its time with its black face uppermost and half with its white, but rather paradoxically this does not mean that the population as a whole will wear an unchanging aspect.

Consider this aspect of the population as a whole. Physically this can be done quite easily by moving far enough away from the table for the individual counters to be indistinguishable. Suppose at the start of the observation that all the counters are arranged to be white side uppermost. Watching from a distance it is noticed that the population loses its full white appearance and becomes darker. It assumes a greyish tone at first pale and then more pronounced; finally it reaches an intensity midway between the white and black. Here however it stops or almost stops; at most a slight oscillation is noticed, at one time to the dark side and at another to the light of this central position. Were the experiment to be repeated with the sole difference that all the counters started off black it would be observed that a similar result supervened; the population would end up with the same mid-grey colour.

What is the reason for this behaviour? The answer is obvious enough, and it can be given in two ways. For any given counter the change from one colour to the other will occur just as quickly (that is, with as little delay) whether it happens for the moment to be white or black. Consequently, if at any instant there happen to be more white counters than black, within the next short interval of time the change white→black will occur more often than the reverse change black→white, and the population will grow darker. Only when each colour is represented in equal numbers will the two directions of change balance, and the general aspect of the population become unchanging. Even then, owing to the fact that a finite number of counters is involved the 'balancing' will not be quite exact, and the overall colour will 'flicker' about the mean.

The second way of looking at things is rather more illuminating. If the observer were asked to arrange the counters so that all were white (or all black) he could do it in only one way. If he were asked to arrange them so that all were white except one, he could do it in one hundred ways; any one of the hundred counters could be the one favoured to be the single black. If he were asked to make all white except two, he could make his choice in $\frac{100.99}{1.2}$ ways; for the first black he could choose any one of the hundred, and for the second any one of the ninety-nine left over. The division by 1.2 is necessary because each one of the combinations would occur twice depending on which of the two counters was chosen first. Similar considerations show that if the observer were asked to leave three counters black he would have a choice of $\frac{100.99.98}{1.2.3}$ ways; and in general if he had to leave r black of $\frac{100.99.98\ldots(100-r+1)}{1.2.3\ldots r}$ ways. It must now be noticed that no matter what actual choice the observer made in fulfilling the commission given him, the result from a distance would look the same. For instance the 4,950 ways of arranging for only two counters to be black would all appear alike, provided they were not observed at close range. In technical language we may say that the 4,950 *microstates* all represent the same *macrostate*; the reader will readily understand this as a shorthand way of saying the same thing. Returning to the model, there is a further observation that can be made; in the process of spontaneous and random jumping the counters fall into every conceivable arrangement at one time or another. They pass through every possible microstate; and, what is significant for our purpose, each microstate is as likely to turn up as any other, or in other words occurs just as frequently. Suppose the integers 1 to 100 are listed and 'black' or 'white' is written arbitrarily against

9. THE STATISTICAL INTERPRETATION OF EQUILIBRIUM AND ENTROPY

each one. Having done this, a microstate has been defined. If this is repeated differently, a new microstate is defined; but in the behaviour of the jumping population each of these microstates is as often realized as the other. It is like the chance of obtaining heads and tails when tossing a coin, only a bit more complicated.

In the model, every microstate belongs to one or other of one hundred and one different macrostates. There is the macrostate in which none of the counters is white; or the macrostate in which only one is white and so on up to the case when all the hundred are white. However, the number of microstates comprising each of these different macrostates varies widely. The first has only one, the second one hundred, the third 4,950 and the $(r+1)$th

$$\frac{100.\ 99.\ 98\ldots(100-r+1)}{1.\ 2.\ 3\ldots r}.$$

Thus in the process of random jumping, while all *microstates* occur just as frequently, some *macrostates* occur very much more frequently, in fact enormously more so. This is already apparent when the first three macrostates, whose relative frequencies are $1:100:4950$, are compared; but when the number of ways for the middle-of-the-range configurations are calculated the figures become quite astronomical, in fact the number for the $50:50$ configuration is over 10^{29}! This is the measure of the proportion of its life in which the population wears, at a distance, the mid-grey aspect. Against these enormous odds it is no wonder that it is never actually observed to revert spontaneously to anything even remotely approaching the pure white or pure black with which it started.

The matter can be summed up by saying that the macroscopic aspect of the population of counters corresponding to a $50:50$ proportion of black and white is the state to which it tends spontaneously; it is the *equilibrium* configuration. It has this property because it is the most probable of all possible configurations; more microstates belong to it than to any other macrostate. However the system starts off, it will move towards this state but never away from it, at least not by more than infinitesimal amounts; but this behaviour is statistical, and not rigidly mechanical.

Two further points remain to be mentioned here. Firstly, the relative preponderance of the middle state over the others increases very sharply as the number of counters goes up (see Fig. 9.1). A population of only one hundred counters has been considered; but real systems which can be handled usually comprise numbers of molecules which run into the million million million range at least. For all practical purposes therefore the equilibrium state is one which is very precisely defined indeed and the spontaneous fluctuations on either side of it (which our model suggests) are quite imperceptible. Only in microscopic systems of the size of bacteria and mitochondria, or in special cases like the critical state, may these fluctuations become significant.

The second point is more mathematical. Suppose the number of microstates corresponding to a given macrostate is called W. Then the logarithm[1] of W is a property of the system that behaves exactly like the entropy. Since the logarithm is a function which always increases with its argument (that is, $\ln x$ always increases with x), $\ln W$ will be a maximum when W is a maximum, and the criterion found for the population of counters is simply

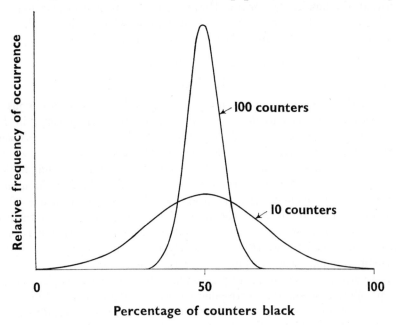

FIG. 9.1. Precision of the equilibrum state with increase of numbers.

this: the equilibrium condition is the state of maximum entropy. Maximum entropy therefore corresponds to maximum probability. Further, if there are two independent populations of counters and a macrostate is settled on for each separately; and if W_1 is the number of possible corresponding microstates for the one and W_2 for the other, the number of microstates for the two together is $W_1 \times W_2$ by a well known law of probabilities, since each microstate of the first can be combined in turn with all the microstates of the second. When logarithms are taken the multiplicative relation ($W_1 \times W_2$) becomes the additive one ($\ln W_1 + \ln W_2$) and this is clearly in agreement with the fact that the total entropy of two systems is the sum of their individual entropies.

[1] Or rather $k \ln W$ where k is Boltzmann's constant, and ln is the natural logarithm.

Widening the Analogy

The model just discussed serves very well for such simple systems as the chemical equilibrium between two optical isomers:

D-isomer ⇌ L-isomer

where the two species have identical energy contents. Or it could serve for a diffusion system in which a solute moves between two identical compartments. It is limited to simple cases such as these because, owing to the symmetry of the counters, there is no energy difference between a counter white-side-up and one black-side-up. It is because of this (which implies that ΔU for the system is zero) that entropy is the sole determining factor in the equilibrium.

Suppose now that the model is made more general by specifying that the centre of gravity (G) of the counters is nearer one face, say the black face (Fig. 9.2). This will mean that as they fall the counters will tend to rotate

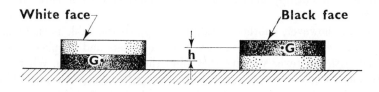

Fig. 9.2

in the resisting air so that the black face is underneath. Two consequences will follow. A counter with the white face uppermost will possess a lower potential energy on coming to rest than one with the black face uppermost, in fact the difference will be mgh, where m is the mass, g stands for the acceleration due to gravity, and h is the distance between the centres of gravity of the two counters. Thus for the whole system of N counters the change from black to white will correspond to a value of ΔU equal to $-N.mgh$. We now have a model for such an isomeric change as

fumaric acid ⇌ maleic acid

or for such a diffusional movement as

aqueous phase $\underset{\rightleftharpoons}{\text{urea}}$ oily phase

in both of which cases the molecule gains or loses potential energy as it moves from one side to the other.

The second consequence is that when a counter lands with its white side uppermost it will do so with a higher velocity than when it lands in the reverse way, for its centre of gravity will have fallen further. If the population of counters shows predominantly white the mean kinetic energy will therefore be higher than if it shows predominantly black. One might say that as the population becomes 'whiter' so its 'temperature' tends to rise. The effect of this on the position of equilibrium will depend on circumstances.

Suppose first that the table on which the counters are jumping is such that it absorbs no energy from them. It will readily be appreciated that this entitles us to regard the counters as an isolated system, for they exchange no energy with their surroundings. This will have the important consequence that the counters which land with the white side uppermost, and therefore possess most kinetic energy on landing, will on that very account jump soonest and so will spend less time on the table.

This effect (the point is illustrative only and must not be pressed too far) will counterbalance the tendency for more to come down this way up, with the result that the equilibrium configuration will again be the 50:50 one. Thus again in an *isolated* system the equilibrium is determined by considerations of entropy alone, and energy relations do not enter into the matter.

There is, however, another possibility. Suppose the table is such that it absorbs an excess of kinetic energy in the counters and supplies a deficit, so that with whatever velocity they land they all jump alike. Whether the system as a whole moves towards 'whiteness' or 'blackness' the mean kinetic energy of the counters remains the same, held constant by interaction with the table. In other words the population is effectively *isothermal*. Counters falling on the table with the white side uppermost will remain there as long as their opposites, and the tendency for more to fall with this orientation will be reflected in a shift towards 'whiteness' of the population equilibrium. How far will it move in this direction? That is impossible to say without further information, but two things are obvious: firstly, that considerations of entropy or probability will always have *some* influence, so that the system never moves completely to the extreme; and secondly, that the more heavily weighted the counters are on one side (i.e. the greater the ΔU) the further will the equilibrium be pushed from the 50:50 configuration. It is these two opposing tendencies of entropy and energy that in the postulated isothermal case are integrated so happily by the free energy function. Thus the model illustrates what was discussed earlier; when the temperature is held constant it is the value of $F = U - TS$ which is the critical factor. As more counters show 'white' the value of the energy U (and so of the free energy F) of the system falls, since its centre of gravity becomes lower; but coincidentally once the white counters outnumber the black ones the entropy begins to decrease (see Fig. 9.2) and the effect of this is to raise F. It then becomes a question of how soon the progressive lowering of S offsets the progressive

9. THE STATISTICAL INTERPRETATION OF EQUILIBRIUM AND ENTROPY

fall in U; ultimately as mentioned it brings the macroscopic movement to an end at the appropriate position of equilibrium.

It is easy to see that the somewhat artificial model we have been discussing has many limitations and inadequacies (one of which is noted below, see footnote[1], p. 121). Of course it proves nothing, as it is only designed to illustrate; but it does bring out in a simple way what has come to be regarded as the real meaning of entropy, and it throws light on the fact that while in one set of circumstances entropy alone dictates equilibrium in another this is determined by the free energy. With the model as a background to our thinking some further aspects of entropy may now be discussed from the statistical point of view.

Factors Influencing Entropy

In thermodynamic systems two entities usually have to be dealt with, matter and energy, and the distribution of both of these has a bearing on the question of the entropy. It must be remembered that the fundamental question is, in how many different ways (that is, microstates) can the material and energy content of the system be arranged so that it looks the same, macroscopically? Confining attention to the material content it can be seen that an *increase in volume*, other things being equal, must mean an increase in entropy. Perhaps this will be clearer if space is regarded as being made up of myriads of minute unit cells; a larger volume then means a greater number of cells in which to distribute the molecules of the system, with an implied increase in the number of microstates yielding any given macrostate. In fact a comparison of the entropies defined classically and statistically for a perfect gas at varying volume is given later, and illustrates this point.

The argument which shows why an increase in volume connotes an increase in entropy can be applied to the case of a mixture which has not yet attained uniformity. In this latter situation the components of the mixture in process of interdiffusion can, in a loose way, be regarded as individually not having availed themselves of the whole volume accessible to them; as they attain equality of concentration everywhere so the effective volume occupied by each component rises, and with it the entropy as defined statistically. This leads to a second consideration; the *process of mixing* (again, other things being equal) implies an increase in entropy. Incidentally, that is why the phenomenon of osmosis (in which mixing is a very prominent feature) can be regarded as primarily an entropic phenomenon (that is, promoted by the tendency of the entropy to increase), while imbibition, another very important biological phenomenon, is a consequence rather of energy relationships. The two phenomena thus reflect the influence of the two terms of the free energy, U and TS.

This brings us to a rather more general point. Entropy can often be helpfully regarded as a *measure of the disorder* in a system (or 'negative entropy' $(-S)$ as a measure of the order). It need hardly be argued that this idea is in agreement with the statistical considerations that have just been discussed; beads on a string are in a much more ordered state, and a far less probable one, than the same beads scattered about the box. This consideration leads at once to some useful conclusions. In a chemical reaction like

$$\text{starch aq.} \rightleftharpoons \text{glucose}$$

the natural tendency is for the reaction to proceed spontaneously far to the right, not because it is highly exothermic (which it is not) but because it involves an increase in disorder or entropy, the glucose units here taking the place of the beads. The same remarks apply to other macromolecules such as proteins, and that is why the problem of the energetics of their synthesis is such an interesting one. Of course it must not be forgotten that sometimes considerations of energy as well as of entropy may play an important part, as in the reaction

$$\text{starch aq.} + \text{inorganic phosphate} \overset{\text{phosphorylase}}{\rightleftharpoons} \text{glucose 1-phosphate}$$

where the equilibrium is on this account, near the centre.

Nor must it be forgotten that as the glucose or other units achieve their freedom from an ordered arrangement so a certain number of water molecules are chemically taken up and find their freedom curtailed, but this is a relatively small effect. What it is necessary to emphasize is that polymerization or condensation into large macromolecules implies a decrease in entropy just because it implies an increase in order; and this increase means that other things being equal the synthesis requires work, whereas the hydrolysis is spontaneous. 'Other things being equal' of course refers primarily to energy, for if there is a large difference in the energy content of the reactants and products (as in the phosphorylase reaction) the effect of the entropy changes may be overshadowed.

The same considerations also apply on what may be called the architectural level in the cell. The ordered structure of the cell membranes, of the chloroplasts and of the chromosomes represent arrangements of low entropy, and that is one reason why growth requires the expenditure of metabolic free energy.

Energy Distribution

Just as the entropy of a state is regarded as reflecting the number of possibilities that exist for arranging the material atoms and molecules spatially within its volume while preserving the identical macroscopic appearance, so the possibilities deriving from the other entity, energy, have

9. THE STATISTICAL INTERPRETATION OF EQUILIBRIUM AND ENTROPY 121

to be considered[1]. Somewhat like matter, energy exists in small discrete units, called quanta, and these have to be distributed not only to individual atoms and molecules, but also within the different 'degrees of freedom' belonging to each of them. Thus, having decided in imagination on a spatial distribution of the molecules we are left with the problem of their energy. The energy of the system can, in thought, be imparted to the molecules in such a fashion as to leave them moving in an essential Brownian fashion; it goes without saying that there will be a vast number of ways (microstates) of doing this all of which will be indistinguishable macroscopically. On the other hand, the energy could be imagined to be imparted in such a way that each molecule moved with the same translational velocity in the same direction (making the whole system into a projectile). Such a pattern of energy distribution would be a fully ordered one, and a moment's thought will convince the reader that there is only one way (microstate) of realizing it.

What these considerations relating to energy amount to is this: whenever a system takes in energy, either as heat or work, its entropy rises in consequence, since with more energy quanta present there are naturally more possibilities for allocating them within the system. Every molecule (on the average) possesses more quanta, and can dispose them about its person in a greater variety of ways, while still maintaining appearances. Here an interesting difference arises, which can be illustrated with reference to volume changes. When the energy is taken in as ordinary *work* there is a decrease in the volume of the system or some part of it, and this decrease in volume *reduces* that component of the entropy which depends on the spatial arrangement of the molecules, as discussed earlier. The result is an *increase* in the number of microstates belonging to the new macrostate because of the presence of more energy quanta, together with a *decrease* in the number as a consequence of there being fewer unit cells of space for the molecules to inhabit. In the special case in which the work is done on the system reversibly the increase from one cause exactly balances the decrease from the other. Inasmuch as a temperature change usually occurs with the doing of work (let the reader think of the compression of a gas) it must be stipulated that the system remains thermally isolated[2], or else some of the energy quanta imparted escape by thermal loss, and the equality noted no longer holds. It is because of this equality, however, that the entropy defined statistically in terms of microstates does not change in a situation (reversible performance of work on a thermally isolated system) *in which the classical definition of entropy also makes it constant*. In short, the reversible performance of work on a thermally isolated system always results in just enough spatial 'ordering'

---

[1] One of the inadequacies of the model discussed earlier arises from its failure to provide for this possibility.

[2] That is, the operation is conducted adiabatically.

of the molecules to exactly off-set the extra 'disorder' introduced by the additional quanta of energy imparted, and it is not difficult to appreciate how the performance of work does this. Volume decrease has been the case discussed, but magnetic or electrical forces clearly act to the same purpose.

The interesting difference mentioned a moment ago appears when, turning from the case of the reversible performance of work, we consider the communication of *heat*. The acquisition of energy by the system again implies an increase in its entropy, statistically defined; but now this increase is not offset by any spatial ordering of the molecules. 'Heat' is itself chaotic in nature, and its acquisition leaves the matter in the system just as spatially disordered as before. These considerations help to link the classical definition of entropy to the statistical definition; for if work reversibly performed on a system cancels out its own effect on the entropy, whereas heat imparted does not, it would seem to be very natural to measure entropy in a process in terms of the heat absorbed. Of course the process must be carried out reversibly, for only then does any work involved cancel out its own effect. Why the absolute temperature also enters into the classical definition of entropy is a point which will not now be discussed.

Appendix I: The Entropy and Volume of a Perfect Gas

As a simple illustration of the equivalence of the classical and statistical definitions of entropy the dependence of the entropy of a perfect gas on its volume will be calculated at constant temperature.

Consider one mole of a perfect gas containing N molecules, N being Avogadro's number. Then with the usual notation the work done by a reversible isothermal expansion from volume V_1 to volume V_2 is given by

$$\text{Work} = \int_{V_1}^{V_2} P \, dV = \int_{V_1}^{V_2} \frac{RT}{V} \, dV$$

$$= RT \ln \frac{V_2}{V_1}. \tag{9.1}$$

Since the internal energy U of the gas has not changed (T having remained constant) this amount of work must have been balanced by heat absorbed from the surroundings. Thus

$$Q_{\text{rev.}} = RT \ln \frac{V_2}{V_1} \tag{9.2}$$

and the change in entropy of the gas becomes

$$\Delta S = Q_{\text{rev.}}/T = R \ln \frac{V_2}{V_1}. \tag{9.3}$$

This is the value of the entropy change as *classically* defined.

We now approach the matter statistically. Imagine the volume V to be divided into a very large number of unit cells, each of volume v. These are extremely small, but nevertheless large enough to contain very many molecules each. These two requirements, very numerous cells each with a high population of molecules, will not be difficult to envisage since there are the enormous number $N = 6\cdot023 \times 10^{23}$ molecules in all. If the gas is internally at rest we may consider that all cells contain the same number of molecules, namely

$$n = \frac{v}{V} N. \tag{9.4}$$

The problem therefore is this: given N molecules, assumed distinguishable[1], in how many different ways can they be arranged in the volume V so that each cell contains the number n? This is in fact a fairly simple problem in permutations and combinations, and the answer is

$$W = \frac{N!}{(n!)^{N/n}}$$

or

$$\ln W = \ln N! - \frac{N}{n} \ln n! \tag{9.5}$$

An approximation now has to be introduced known as Stirling's formula, but since it is highly exact under conditions which the Second Law itself presupposes (that is, that very large populations are involved) the fact that it is an approximation need not concern us. Stirling's formula is

$$\ln N! = N \ln N - N. \tag{9.6}$$

That this formula is approximately true can be seen by considering the area under the curve $y = \ln x$ from $x = 1$ to $x = N$. This area is equal to

$$A = \int_1^N \ln x \, dx = \left[x \ln x - x \right]_1^N$$
$$= N \ln N - N + 1$$
$$= N \ln N - N \quad \text{very nearly}. \tag{9.7}$$

Referring to the figure (Fig. 9.3), an approximation to the area can be made by regarding each of the intervals shown as trapeziums and summing them. This gives the result

$$A = \tfrac{1}{2} \ln 1 + \ln 2 + \ln 3 + \ldots + \ln(N-1) + \tfrac{1}{2} \ln N$$
$$= \ln N! - \tfrac{1}{2} \ln N. \tag{9.8}$$

Since N is very much greater than $\tfrac{1}{2} \ln N$ (the reader can easily verify this) equating the two expressions for the area yields Stirling's formula.

---

[1] Modern physical theory denies that they can be distinguishable, but this point need not concern us here.

We now substitute in equation (9.5) which gives

$$\ln W = N \ln N - N - \frac{N}{n}\left(n \ln n - n\right)$$
$$= N \ln N - N \ln n$$
$$= N \ln \left(\frac{N}{n}\right). \tag{9.9}$$

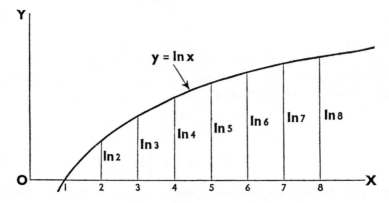

Fig. 9.3. Derivation of Stirling's approximation.

Recalling that $\dfrac{N}{n} = \dfrac{V}{v}$, the equation can be written out twice using subscripts to refer to two volumes V_1 and V_2:

$$\ln W_1 = N \ln \left(\frac{V_1}{v}\right)$$
$$\ln W_2 = N \ln \left(\frac{V_2}{v}\right). \tag{9.10}$$

Subtracting, and multiplying by $k = \dfrac{R}{N}$ to make the right hand side identical with that of equation (9.3) gives the result

$$k \ln W_2 - k \ln W_1 = R \ln \left(\frac{V_2}{V_1}\right) \tag{9.11}$$

where R is the gas constant and k is Boltzmann's constant.

Thus this simple calculation illustrates what was said before, that the function $k \ln W$ behaves like the classically-defined entropy. It should be noted that provided only *changes* in entropy are considered the size of the unit cells (v) disappears from the final expression, and it is therefore immaterial.

However, if the absolute value of the entropy is to be calculated then the value of v must be known. It is this further information which, in effect, the quantum theory is able to give.

Appendix II: The Entropy Change on Melting

It is instructive to consider in the light of the previous discussion the entropy change which accompanies the melting of a crystalline solid like ice. The change is given by classical thermodynamics as

$$\Delta S \text{ (melting)} = L/T_m$$

where L is the latent heat and T_m the melting point. Two comments can be made on this from the statistical point of view. Firstly, there is clearly a *marked increase in disorder* when the regular crystal lattice becomes the chaotic molecular configuration of the liquid. This bears out the idea that entropy reflects the degree of disorder.

Before making the second comment it is neceessary to say something in an elementary way about the Principle of the Equipartition of Energy. Consider a monatomic gas like helium; in order to specify what has been called the thermal energy of a molecule three independent components would have to be named, the kinetic energies parallel to three rectangular axes. These might be called (see Fig. 9.4)

$$\tfrac{1}{2} mv_x^2, \ \tfrac{1}{2}mv_y^2, \ \tfrac{1}{2}mv_z^2$$

and they represent the only energy possessed by the molecule which is subject to the 'shuffling around' processes associated with thermal agitation. (The latent heat which is given out on condensation represents potential energy which lies outside this process.) Now consider a diatomic molecule. Besides the three components already mentioned there are two more associated with rotation, and two more associated with vibration. These can be written

$$\tfrac{1}{2}I\omega_1^2, \ \tfrac{1}{2}I\omega_2^2, \ \tfrac{1}{2}mv_d^2, \ \tfrac{1}{2}\epsilon d^2$$

where I is a moment of inertia; ω_1, ω_2 are angular velocities; v_d is a vibrational velocity; ϵ is an elastic constant and d is the excess distance between the atoms over the value at rest. The details of these terms are not so important here however; what matters is their existence. The reader may wonder why only two axes of rotation are concerned, and not also the long axis joining the atoms. The answer is that rotation about this latter axis, owing to the symmetry of the molecule, is not excited (or otherwise) by molecular collisions; consequently its energy does not count as thermal energy and take part in the general randomization. Thus while it can be said that a monatomic gas possesses three degrees of thermal freedom, a diatomic gas possesses seven. Larger molecules would of course possess more.

We now consider the Principle of the Equipartition of Energy. Except where quantum effects become important it can be shown that at statistical equilibrium the average energy associated with each of these 'square terms' (as they are called) is the same, and is equal to $\frac{1}{2}kT$ per molecule or $\frac{1}{2}RT$ per mole. Consider in the light of this the heat capacity per mole of a pure substance. This will obviously be proportional to the number of the

Monatomic gas

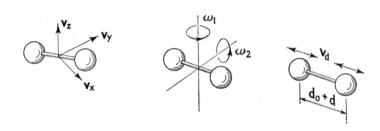

Diatomic gas

FIG. 9.4 Degrees of freedom of monatomic and diatomic gas molecules.

degrees of thermal freedom of its molecules, since as the temperature is raised, enough heat must be supplied to keep each of them 'topped up', as it were, to the value $\frac{1}{2}RT$. That is why a diatomic gas has a higher specific heat than a monatomic one. In the case of melting ice the principle is involved in the same way. When the molecules break free from the crystal lattice and acquire new degrees of thermal freedom (that is, when the ice changes to water) the specific heat shows a sudden rise to nearly twice its previous value. But the acquisition of the new degrees of freedom is also the underlying cause of the phenomenon of the latent heat, since as these new degrees become available each has to be filled up with energy to the level $\frac{1}{2}RT$ before any general rise of temperature is possible. The situation is illustrated in Fig. 9.5 by a simple analogy. The vertical tubes represent the degrees of thermal freedom of the molecules, and the total volume of the liquid in them represents the internal energy U of the system. As the temperature T rises so the volume of contained liquid rises by flow through the lateral holes at the base of those tubes representing the 'accessible' degrees of freedom. At T_m (the melting point) some new tubes become accessible, and the liquid level (that is T) remains stationary while these fill up with a volume corresponding to the latent heat. This volume is proportional to (1) the number of new tubes, and (2) the height T_m. Thus if the volume (representing L) is divided by T_m the quotient measures the number of new tubes which become accessible at the melting point.

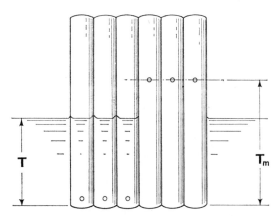

Fig. 9.5. Acquisition of new degrees of freedom on melting.

Thus the Principle of Equipartition makes it easier to see how the acquisition by the molecules of new degrees of freedom (a phenomenon of very obvious importance to the whole idea of statistical microstates) links up with the phenomenon of latent heat (which is of equally obvious importance in the classical idea of the entropy of melting). The model that has been discussed even suggests why it is that in the classical approach the quantity of heat involved (L) must be divided by the absolute temperature (T_m); the quotient is a measure of the number of extra degrees of molecular freedom realized on melting.

CHAPTER 10

Chemical Reactions and Membrane Equilibria

'Well?' said Rabbit.
'Yes,' said Owl, looking Wise and Thoughtful. 'I see what you mean. Undoubtedly.'
'Well?'
'Exactly,' said Owl. 'Precisely.'
'Why?' asked Rabbit.
'For that very reason,' said Owl, hoping that something helpful would happen soon.

THE HOUSE AT POOH CORNER
A. A. Milne

So far the discussion of equilibrium has been concerned exclusively with closed systems, that is, with definite fixed portions of matter. For many of the physical applications of thermodynamics this is not a serious limitation, but for biologists it comes a long way short of covering all the cases of interest. In the first place, biological systems are far from being closed. One can think of an organelle like the mitochondrion or chloroplast ceaselessly exchanging chemical substances with its cytoplasmic environment; of the cell itself across whose boundaries there is a constant flux of water, minerals and organic solutes; or of the entire plant as the seat of the processes of growth and development. In the second place, the plant physiologist is peculiarly interested in metabolism, the whole complex of biochemical activity within the plant. While many chemical changes can take place in closed systems and do not necessarily involve the idea of openness, yet in an important sense they are always linked with it. In an open system molecular species enter or leave by transport processes across boundaries, whereas in chemical processes (in a closed system) they may also be considered to enter or leave, even if they do so through what might be called a 'chemical back door'. The main difference from the thermodynamic point of view is that in purely chemical processes the 'entry' of one or more chemical species necessarily involves the 'exit' of others, since mass is conserved. Further, the amounts of different species involved are fixed in their proportions and these requirements naturally impose restrictions which are not usually present in the transport case. Still, the fact that exit and entry (in the wider sense of the terms) is involved in both chemical changes and open systems gives the two cases important points of similarity for thermodynamics.

The Chemical Potential

This discussion can be started by considering a situation of frequent interest to the physiologist. Imagine it in its simplest form, a system consisting of just two phases which will be called α and β (Fig. 10.1). (A phase, incidentally, is a portion of a system characterized by being homogeneous and by having a definite boundary.) Suppose further that each phase contains two components, an unspecified solute (1) and solvent water (2). The problem is this:

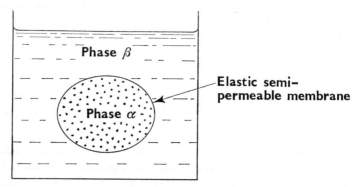

Fig. 10.1. Solvent equilibrium between phases.

what property of the water determines whether it moves from α to β or vice versa, assuming that movement is not prevented? The reader will appreciate that if the boundary between the phases has special properties like elasticity and semi-permeability (and the physiologist often meets with such boundaries) then there is no immediate and easy answer to the problem; the water (and solute) will not simply move till *equality of concentration* has been achieved. So to answer the problem generally, some more subtle measure is required.

This simple example clearly embodies the essence of the problem of the open system, for each phase is in effect an open system capable of exchanging matter with its surroundings, represented by the other phase. Therefore, if this problem can be solved the general one has also been solved; the fact that considerations have been confined to a system of no more than two phases and two components, and moreover that only the movement of water has been discussed is neither here nor there.

The first point to notice is that *there must exist* some measure which will determine whether a substance will move or not from one system or phase to another. The position is rather similar to that which gives rise to the idea of temperature. Two bodies in thermal equilibrium with a third are found to be in equilibrium with one another, and that makes 'temperature' an unequivocal thing; it is not nonsense to speak of a body as possessing a 'temperature'. Similarly, two phases both in equilibrium with a third as regards gain or loss of water are found when placed in suitable contact to be in equilibrium with one another. One may think of potato tissue in equilibrium with a sucrose solution; it neither gains nor loses water when transferred to the closed, steady atmosphere above the solution. It is this general sort of situation that makes it possible to speak of the *chemical potential* of the water (or of any other species) just as in the analogous case we can speak of the temperature of a body. The chemical potential of a substance dictates the direction of its spontaneous movement rather as temperature dictates that of heat flow.

The second point to notice is that it must be possible so to define the chemical potential that if unit quantity of matter passes from one potential to another the difference in potential is equal to the maximum amount of work[1] which is involved. Lewis's statement of the Second Law implies that a spontaneous movement between phases or systems is capable of doing work. This work is obviously proportional to the amount of matter which passes. This suggests that we define the chemical potential in such a way that when the difference in potential, say $(\mu^\alpha - \mu^\beta)$, is multiplied by the amount of matter which passes (dn moles) the result is the maximum amount of work which could be gained from the movement if suitably harnessed. In symbols,

$$\text{maximum work} = (\mu^\alpha - \mu^\beta)\, dn. \tag{10.1}$$

Perhaps in connection with this procedure it is worth reminding the reader firstly, that when temperature has been established as an unambiguous concept it is still open to us to choose any sort of scale we please, and in fact the one we choose fits in neatly with ideas of work[2]; and secondly, that the definition of chemical potential runs closely parallel to that of electrical potential (measured in volts). Reference to an elementary textbook of electricity will confirm this, though it will appear later that the thermodynamic conception is more general in its scope.

RELATION OF CHEMICAL POTENTIAL TO THE FREE ENERGY

It now remains to connect the new idea of chemical potential with the previous ones, in particular with the two free energies. The connection in fact follows very readily from equation (10.1), but before it is derived an important point must be made explicit. To speak simply of transferring a quantity (dn moles) of substance from one phase (α) to another (β) is not

---

[1] This presupposes equality of temperature, a point taken up later. Where there is a difference of temperature there *is* no unique maximum to the work obtainable. It then depends on the whole 'pathway' along which the movement takes place.

[2] The reader may care to consult Kelvin's original definition of the Absolute Scale in any elementary textbook on Heat.

really precise enough. As soon as matter is removed from phase α something will happen to the thermodynamic variables of this phase, such as its volume, pressure and entropy. If we proceed in a casual way it may well be that all three of these variables wander, and in a quite undefined manner; obviously the reverse will happen to phase β when the dn moles are added to it. It is not a question of whether all the operations are conducted reversibly or not; rather it is a question of how the phases are 'held' mechanically and thermally by their surroundings, and of the sort of experimental arrangements that are provided. It is no wonder, therefore, that if precise relationships are to emerge the specification as to the conditions attached to the imagined transfer must be precise. The situation is, in fact, rather similar to that which was faced in discussing virtual displacements as a test for thermodynamic equilibrium.

There are two sets of conditions, in particular, which are especially important. This is because they are realistic in the sense that not only can they be easily realized in experimental work, but they also correspond very closely to the state of affairs which occurs in nature. They can be expressed in this way: firstly, the transfer of dn moles of substance is imagined to take place under conditions in which the temperatures of the two phases remain equal and constant, and their *volumes* (V^α and V^β) remain unchanged; and secondly, under conditions in which the temperatures being again equal and fixed the *pressures* (P^α and P^β) remain unchanged (though not necessarily equal). Of these two sets of conditions the second is very much more important to the physiologist. These sets of conditions can be summarized as follows:

Set (i): $T^\alpha = T^\beta$ fixed; V^α, V^β fixed; P^α, P^β free to alter.

Set (ii): $T^\alpha = T^\beta$ fixed; P^α, P^β fixed; V^α, V^β free to alter.

Consider these conditions in turn. Suppose a quantity dn_1 moles of component (1) is transferred by any suitable reversible process[1] from phase α to phase β under the set of conditions (i). Then the work obtained (it will necessarily be the maximum, as the operation is reversible) will be equal to the total decrease in the Helmholtz free energy of the system. But this is made up of two parts, the decrease associated with phase α and the increase associated with phase β. Mathematically these two components can be written $\dfrac{\partial F^\alpha}{\partial n_1}\,\mathrm{d}n_1$ and $\dfrac{\partial F^\beta}{\partial n_1}\,\mathrm{d}n_1$. However, it has been decided that the work shall be expressible by the product $(\mu_1^\alpha - \mu_1^\beta)\mathrm{d}n_1$. We therefore have

$$(\mu_1^\alpha - \mu_1^\beta)\,\mathrm{d}n_1 = \frac{\partial F^\alpha}{\partial n_1}\,\mathrm{d}n_1 - \frac{\partial F^\beta}{\partial n_1}\,\mathrm{d}n_1. \qquad (10.2)$$

---

[1] It is very helpful in fixing things firmly in the mind to think out, for a simple case, what is meant by a 'suitable reversible process'. For two examples refer back to Chapter 7.

10. CHEMICAL REACTIONS AND MEMBRANE EQUILIBRIA

Since this is to be identically true, and μ_1^α and μ_1^β are independent then we must have separately

$$\mu_1^\alpha = \left(\frac{\partial F^\alpha}{\partial n_1}\right)_{T, V^\alpha, n_2}$$
$$\mu_1^\beta = \left(\frac{\partial F^\beta}{\partial n_1}\right)_{T, V^\beta, n_2} \qquad (10.3)$$

where the partial derivatives have been written out in full. It will be readily seen by the reader that this line of approach leads at once to a general definition of the chemical potential μ_i of a species i in a system of any number of components:

$$\mu_i = \left(\frac{\partial F}{\partial n_i}\right)_{T, V, n_j} \qquad (10.4)$$

where the subscript n_j means that the amounts of all the components except the one being considered (i) are held constant. Sometimes a little rider ($j \neq i$) is added to the equation to indicate this last point, but it is hardly necessary. This then is the relationship that is required.

RELATION TO GIBBS FREE ENERGY

Now that a definition for μ has in effect been found in terms of the Helmholtz free energy its relation to the Gibbs free energy must be considered. This is a very simple matter to decide. Suppose that dn_1 moles have been transferred under the previous conditions (constant T, V^α, V^β) and that the work released is safely stored up in whatever form we wish, say as the energy of a coiled spring. Next, imagine the constraints that held the volumes constant to be loosened in such a way that the phases can expand or contract enough to re-establish their original pressures. In this readjustment some more work will be done, positive or negative, and this will naturally be reflected in a further change in the total Helmholtz free energy. However, if the reader will reflect for a moment he will realize that this further work adds nothing to the energy of the spring; it is all associated with volume changes at constant pressure[1]. The energy in the coiled spring remains at the end of the pressure readjustment as the *useful* work associated with the transfer of the dn_1 moles, and as such it is equal of course to the overall decrease in G. Thus we may write

$$(\mu_1^\alpha - \mu_1^\beta)\, dn_1 = \frac{\partial G^\alpha}{\partial n_1}\, dn_1 - \frac{\partial G^\beta}{\partial n_1}\, dn_1 \qquad (10.5)$$

---

[1] It might be objected that the pressure has not remained constant. However, it has only varied infinitesimally, and so can be considered as constant. The 'error' in this is of the second order of small quantities, and so ultimately becomes vanishingly small.

where the partial derivatives are now taken with the pressures constant instead of the volumes. This in turn leads to the general definition of μ_i as

$$\mu_i = \left(\frac{\partial G}{\partial n_i}\right)_{T,P,n_j} \tag{10.6}$$

and this is an alternative to (10.4). As a matter of fact, in the bewildering way which its many variables make possible, thermodynamics offers still other definitions of μ. Gibbs originally defined it by the relation

$$\mu_i = \left(\frac{\partial U}{\partial n_i}\right)_{S,V,n_j}, \tag{10.7}$$

and still another definition is

$$\mu_i = \left(\frac{\partial H}{\partial n_i}\right)_{S,P,n_j}; \tag{10.8}$$

but beyond knowing that such definitions exist (in case he comes across them) the physiologist need not concern himself with them. They are certainly far less easily visualized in terms of practical laboratory operations than the first two.

Condition of Equilibrium

When a process is such that work can be obtained from it, that process will occur spontaneously, if it is allowed to. This is in fact one way of stating the Second Law, since if the spontaneous direction happened to be the reverse one, then the situation would arise that ability to do work could be generated spontaneously, which unfortunately is not the case. If the reader reflects he will see that this implies that a chemical substance will always move from a place where its potential (μ^α) is higher to one where it is lower (μ^β) since in this movement work proportional to the change ($\mu^\alpha - \mu^\beta$) can be obtained. Movement naturally ceases when the two potentials are equal. This at once gives the very important result that the condition of transport equilibrium between two phases is simply

$$\mu^\alpha = \mu^\beta \tag{10.9}$$

and nothing more need be added[1]. Perhaps it should be mentioned however that the condition of equilibrium is often stated a trifle differently. It is said to depend on equality of *activities* in the two phases, rather than on equality of chemical potentials. However, no contradiction of what has just been said is implied, for the activity of a substance is simply an exponential function

---

[1] It is implicit that there is uniformity of temperature (though not of pressure); since if matter of any sort can pass so also can heat, and inequality of temperature will therefore imply lack of general equilibrium.

of its potential, and equality of one means equality of the other. In symbols, and in the case of a solution, if a_i is the activity of the solute i then as a matter of definition

$$a_i = \exp\left(\frac{\mu_i - \mu_i^0}{RT}\right)$$

or

$$\mu_i - \mu_i^0 = RT \ln a_i \tag{10.10}$$

where μ_i^0 is a constant for a given solute fixed arbitrarily for convenience in a way to be discussed later. One of the advantages of the activity is that it ordinarily runs fairly parallel with the concentration; provided the solvent does not change it can be looked upon as a sort of 'corrected' concentration. This should be fairly obvious to the reader when he recalls that there is a *logarithmic* relationship also between the work or free energy and concentration in the osmotic dilution of ideal solutions.

OTHER PROPERTIES OF THE CHEMICAL POTENTIAL

There are several other important properties belonging to the chemical potential about which we must be clear. In the first place, its numerical magnitude is undefined to the extent of an additive constant. This is a situation similar to that which holds for so many other important thermodynamic variables (U, S, F and G); only *differences* in chemical potential are defined and that means that a zero can be fixed arbitrarily. This may seem a little paradoxical to the reader, for μ can be defined as a differential coefficient, and this means that its value depends on the *change* (dG) in a thermodynamic quantity (G) whose changes have presumably a quite definite value. The explanation of this paradox is that changes in G are indeed unambiguously defined, but only for *closed* systems (one thinks of the discussion leading to the establishment of U and S, which are components of G). What is done when defining μ is to imagine that new matter (dn) is added to the system, and the new matter brings with it new free energy (dG) whose absolute magnitude is clearly undefined.[1] Hence the indefiniteness is transferred to μ and its zero can be fixed arbitrarily if necessary.

In the second place, the chemical potential is an intensive property, like temperature, pressure or concentration. This is suggested by the fact that it is the ratio of increments of two extensive quantities (dG and dn), and it can be seen very clearly by considering a simple situation. If a complex system of many components and perhaps many phases as well is placed in a closed

---

[1] The free energy dG is not simply that belonging to the matter dn. It is also associated with the *process of introduction* of this matter into the system. However, this point need not be laboured.

container separated from a vacuous space by a membrane permeable only to one of its components (i), then when equilibrium has been reached the chemical potential of i will be the same in all phases including the previously empty space. Hence

$$\mu_i^\alpha = \mu_i^\beta = \mu_i^\gamma = \ldots = \mu_i^\nu \qquad (10.11)$$

where ν stands for the space. A very simple situation of this sort can easily be encountered by the physiologist; in fact all he has to do is to enclose a leaf in a stoppered bottle. The leaf has very many phases, and naturally the number of chemical substances present is very large. However, the only one which can migrate into the empty space in appreciable quantities is water. At equilibrium (barring active mechanisms) the water will have the same potential everywhere, in the chloroplasts, in the vacuole, in the cell walls and in the aerial phase. The latter may be considered as pure water vapour (if the air were removed it would make virtually no difference to the system) and for any *pure* substance the expression $\left(\dfrac{\partial G}{\partial n}\right)_{T,P}$ (there are now no n_j's to keep constant) reduces simply to $\dfrac{G}{n}$, where G is the total free energy of the phase and n the total number of moles of the substance which it contains. Thus, since the chemical potential of the water in the aerial space is an intensive property of that phase[1], so also must it be for the other phases by virtue of equation (10.11).

In the third place, it is obvious that the chemical potential can be regarded as one of the very important class of partial molar quantities (see Chapter 3). It is, in fact, the partial molar Gibbs free energy. The reason, incidentally, why it stands in this special relationship to G (when it can also be defined in terms of U, H, or F) is that in this case the thermodynamic properties which are held constant in the differentiation (that is T, P) are both *intensive* properties. Intensive properties concern the quality of a system, extensive properties its quantity; hence when attention is fixed on the relationship (10.6) between μ and G the system can be considered as being *qualitatively* changed as regards composition only, without dragging in other qualitative changes like alterations of temperature and pressure. When the alternative definitions of μ are employed this simplification is absent. For present purposes the importance of this recognition lies in the fact that the equation follows at once (see Chapter 3)

$$\begin{aligned} G &= n_1\mu_1 + n_2\mu_2 + n_3\mu_3 + \ldots \\ &= \Sigma\, n_i\mu_i \end{aligned} \qquad (10.12)$$

---

[1] That is, if the volume of the space were enlarged it would maintain the same value, since both G and n would increase in the same ratio.

where G is the total free energy of the system. The previous paragraph is really an explanation of the reason why G appears on the left hand side here and not U, H or F, although μ is apparently related just as symmetrically to these as it is to G (see equations (10.4), (10.6), (10.7) and (10.8)). In fact the symmetry is not quite perfect, for only in G are the two variables (T, P), which are held constant, both intensive or qualitative variables. That accounts for its privileged position in equation (10.12).

There is one final point which needs to be made clear, and this is that while such a derivative as $\left(\dfrac{\partial G}{\partial n_i}\right)$ is mathematically quite unexceptionable, thermodynamically it is at first sight a little problematical. The trouble is that the quantities U, H, F and G were all defined for *closed* systems, and that now they are being used in connection with a type of process (addition of matter) which was never originally envisaged at all. A fresh appeal to the facts of nature must be made to justify this; and justification comes, as has already been hinted, by considering the state of affairs in a phase separated from the system by a membrane permeable to just one component (i) and containing nothing but that component. For this pure phase, μ_i has the same value as in all the mixed phases; but in this one case the difficulty mentioned does not arise, for the expression $\dfrac{\partial G}{\partial n_i}$ (which implies adding substance i) reduces simply to G/n (which implies no such operation). But the reader must be left to think this over.

General Equation for an Open System

When a closed system is in equilibrium it has already been found (Chapter 8) that any small displacement in its condition can be described by four equivalent relations (equations (8.4), (8.14), (8.15), and (8.16)). Here we select the equation which is, for the physiologist, the most important:

$$\mathrm{d}G = V\mathrm{d}P - S\mathrm{d}T \quad \text{(closed system in equilibrium)}. \tag{10.13}$$

Suppose to such a system matter is added, regarding it as an open system. If a component i is added in such a way that the temperature and pressure remain everywhere unchanged there will be a further increase in the total Gibbs free energy given by $\left(\dfrac{\partial G}{\partial n_i}\right)\mathrm{d}n_i$, where $\mathrm{d}n_i$ is the amount added. Equation (10.13) therefore becomes, for the case where the change in the system includes gain or loss of matter,

$$\begin{aligned}\mathrm{d}G &= V\mathrm{d}P - S\mathrm{d}T + \left(\dfrac{\partial G}{\partial n_i}\right)\mathrm{d}n_i \\ &= V\mathrm{d}P - S\mathrm{d}T + \mu_i \mathrm{d}n_i,\end{aligned} \tag{10.14}$$

or where a number of components are added simultaneously,

$$\begin{aligned}dG &= VdP - SdT + \mu_1 dn_1 + \mu_2 dn_2 + \ldots \\ &= VdP - SdT + \Sigma \mu_i dn_i.\end{aligned} \quad (10.15)$$

This extension of equation (10.13), it should be emphasized, is in the first place just a pure piece of mathematics. If the definition of μ_i is accepted as $\left(\dfrac{\partial G}{\partial n_i}\right)_{T,P,n_j}$ then, merely by this definition, the addition of substance dn_i increases the total free energy by $\mu_i dn_i$. Thermodynamics comes in because the original equation (10.13) is not just a mathematical identity; and also because a further meaning can be given to μ_i (as was seen earlier). It is left as an exercise to the reader to show that by using the identities (8.11), (8.12) and (8.13) of Chapter 8 the dG of equation (10.15) can be replaced with dU, dH and dF. For dU the result is

$$dU = TdS - PdV + \Sigma \mu_i dn_i. \quad (10.16)$$

Putting dS, dV and all the dn's except dn_i equal to zero we get the result

$$dU = \mu_i dn_i$$

or

$$\mu_i = \left(\frac{\partial U}{\partial n_i}\right)_{S,V,nj} \quad (10.17)$$

which is the relation (10.7) quoted earlier. The other formulae for μ_i follow equally easily.

Chemical Reactions

The conception of a chemical potential, which determines the direction of movement of a chemical substance just as temperature determines that of heat and electrical potential ordinarily determines that of electricity, has very naturally arisen and been established in connection with transport processes. One of the fundamental observations used to establish the notion was that if two phases (say α and β) are each in equilibrium with a third (γ) as regards exchanging a chemical substance i with it, then they will be found to be in equilibrium with one another as regards movement of i (see Fig. 10.2). This really follows from the Second Law, for if α and β are not in equilibrium some of substance i will pass from one to the other. This will presumably disturb phases α and β in different directions, and one will proceed to gain substance i from phase γ while the other will proceed to lose to it. A macroscopic circulation of i will thus be set up spontaneously, a genuine sort of perpetual motion, which is believed to be impossible.

This sort of argument is, however, difficult to apply to chemical reactions if only for the reason that single substances can never be considered in isolation. Even in the simplest chemical reaction

$$A \rightarrow B$$

there are two substances involved, and in the general case an indefinite number have to be allowed for. However, it is fortunate that having worked things out for the case of transport phenomena conclusions can be validated for chemical reactions in a fairly simple way.

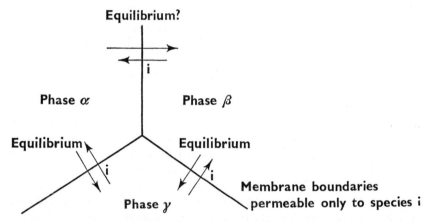

FIG. 10.2. Transport equilibrium between three phases.

Consider firstly in an abstract way what the general equation (10.15) means:

$$dG = VdP - SdT + \Sigma \mu_i dn_i. \qquad (10.18)$$

Consider a system which is in internal equilibrium (otherwise strictly speaking we cannot define its properties thermodynamically). Its free energy is measured, and then a number of operations are conducted upon it. Firstly the temperature of the system is raised by an amount dT; then the pressure on it is increased by dP. Of course, it goes without saying that both these operations must be arranged to leave the system in equilibrium[1], or else it cannot be said that it has a new, well-defined, free energy; and the same proviso applies to the next operations, the addition of dn_1 moles of component (1), dn_2 moles of component (2) and so on. How can a substance be added to a system so as not to destroy its internal equilibrium? Obviously it cannot just be thought of as being poured into the system at one point, in the way

[1] That is, the temperature must still be uniform everywhere, and the pressure uniform within each phase.

that sugar is added to a cup of tea. This would be followed by the spontaneous process of diffusion, with its implied absence of internal equilibrium. What has to be thought of instead is the substance being *increased in concentration everywhere in the system* in the same ratio, $\left(\dfrac{n_i + \mathrm{d}n_i}{n_i}\right)$. If this ratio differs only infinitesimally from unity then the system will still be in internal equilibrium after the addition, which is the required result. Having achieved this state of affairs the free energy is remeasured (in thought) and the increase (dG) is found. Equation (10.18) then asserts that this increase will be related to dT, dP and the dn's in the way stated. Of course, if there was a simple instrument for measuring G and the aim was to test the equation in the laboratory instead of engaging in abstract thinking, there would be no objection to actually adding the constituents in bulk form at isolated points, and in fact this is how it would probably be done. What the foregoing discussion implies is that the free energy of the system should not be remeasured (with a view to calculating its increase) immediately after the addition; an interval must be allowed for the material added to diffuse throughout the system to give the same minute proportional increase in concentration everywhere which was spoken about earlier. Only then, if the free energy increase is measured, could it be expected to fit equation (10.18).

The point of the rather lengthy discussion on this equation now becomes apparent. The end result of adding dn_i moles of substance i to the system is that it is distributed throughout the system uniformly within each phase. *It has not obviously been brought in by a transport process*; it might equally well have entered by a chemical 'back-door'. Consequently there is every justification for considering that equation (10.18) applies to a system with chemical reactions with just as much validity as it does to a system with transport. In fact, the reader will appreciate at once that the outcome of a chemical reaction in a system can be exactly duplicated (so far as the system is concerned) by using only transport processes. If it is required, for instance, to convert some of the maltose in a sample of solution to glucose some maltase can be introduced for a time sufficient for requirements and then withdrawn. Alternatively, some maltose can be dialysed out via one selectively permeable membrane and the necessary amount of glucose dialysed in through another. If equation (10.18) expresses the free energy change of the system in the second case, it must obviously do so in the first as well since the end results are indistinguishable. Incidentally, this example illustrates a further point. Before the free energy of the system is measured it must be ensured that it is in internal equilibrium. How is this done in the presence of a chemical reaction which has not gone to completion? The answer is that the experimental fact is used that very many reactions proceed at a negligible rate in the absence of a catalyst or enzyme, and that enzymes and catalysts

10. CHEMICAL REACTIONS AND MEMBRANE EQUILIBRIA

can be readily poisoned or inhibited. Thus, just as it is possible to start and stop the addition (by transport) of a component to a system by opening or closing a tap or shutter, so a chemical reaction can be started or stopped by adding or poisoning a catalyst. Further, in the same way that the transport of one component can be 'organised' in the presence of others by the use of selectively permeable membranes, so chemical changes can be organised along a desired pathway when there are alternatives, by the use of specific enzymes. All this gives point to the assertion that equation (10.18) is equally applicable to chemical systems, and so we shall proceed to use it to discover the general condition for chemical equilibrium in terms of chemical potentials.

Condition for Chemical Equilibrium

For the sake of concreteness consider a simple chemical reaction into which four substances enter:

$$A + B \rightleftharpoons C + D$$

As written, this equation is only qualitative; to make it quantitative the stoichiometrical numbers ν_A, ν_B, ν_C, ν_D need to be added to 'balance' it:

$$\nu_A A + \nu_B B \rightleftharpoons \nu_C C + \nu_D D. \tag{10.19}$$

Unlike most transport systems the amounts of the substances taking part in a chemical system are not independent of one another. In the above example once the amount (dn_A) of component A which is to be added or removed chemically has been settled, the amounts $(dn_B, dn_C$ and $dn_D)$ of all the others have also been settled. In fact the equations hold:

$$-\frac{dn_A}{\nu_A} = -\frac{dn_B}{\nu_B} = \frac{dn_C}{\nu_C} = \frac{dn_D}{\nu_D} \ (= d\xi \text{ say}) \tag{10.20}$$

where the negative signs appear for obvious reasons. For a single chemical reaction therefore there is really only one quantity which can be imagined to vary independently. It can be regarded as dn_A if necessary; but for the sake of symmetry and mathematical simplicity we choose rather the quantity $d\xi$ which has been given the form of a differential (when it might have been called 'λ' or 'k') since this reminds us that we are dealing in *changes* and that these are to be regarded as being very small.

Now return to the consideration of equation (10.18). Corresponding most closely to general laboratory or physiological situations, imagine the chemical reaction (10.19) as taking place under conditions of constant temperature and pressure. For any small change therefore, both dT and dP can be set equal to zero. This reduces equation (10.18) to

$$dG = \mu_A dn_A + \mu_B dn_B + \mu_C dn_C + \mu_D dn_D \quad \text{(constant } T,P\text{)} \tag{10.21}$$

for the reaction under consideration. Now write $dn_A = -\nu_A d\xi$ and so on from (10.20) for the other quantities. Dividing through by the common factor $-d\xi$ gives

$$-\left(\frac{\partial G}{\partial \xi}\right)_{T,P} = \nu_A \mu_A + \nu_B \mu_B - \nu_C \mu_C - \nu_D \mu_D \qquad (10.22)$$

and this relationship gives at once the required condition. For the reaction to be at its point of equilibrium dG (or $\dfrac{\partial G}{\partial \xi}$) must be zero and this gives the criterion

$$\nu_A \mu_A + \nu_B \mu_B - \nu_C \mu_C - \nu_D \mu_D = 0. \qquad (10.23)$$

Before this is considered more fully however the meaning of the quantity ξ and its differential $d\xi$ can be made clearer. To do this consider the analogy between a simple transport process and a chemical reaction. Crossing a phase boundary in the former case corresponds to crossing from one side of the chemical equation to the other in the latter. For transport processes the condition of equilibrium is

$$\mu_i^\alpha = \mu_i^\beta \qquad (10.24)$$

and for the chemical reaction the condition that has just been found can be written

$$\nu_A \mu_A + \nu_B \mu_B = \nu_C \mu_C + \nu_D \mu_D \qquad (10.25)$$

which is very similar to equation (10.24) but just a little more generalized. Now the analogy can be developed by defining $d\xi$ for the transport case; if dn_i moles cross from one side of the phase boundary to the other then (cf. equation (10.20) above)

$$-\frac{dn_i^\alpha}{\text{unity}} = \frac{dn_i^\beta}{\text{unity}} = d\xi \qquad (10.26)$$

since the numbers which correspond to the stoichiometrical coefficients here are each unity. Thus $d\xi$ is simply the amount of substance in moles (i.e. dn_i) which has crossed the boundary, and an increase in ξ simply measures the 'progress of the reaction' (the transfer of species i from one phase to the other). It is not surprising therefore to find that the interpretation of ξ is very similar for the chemical situation. It measures the 'extent of the reaction' and $d\xi$ can be looked upon as the amount of matter, in moles, which crosses from one side of the reaction equation to the other. Only in calculating the number of moles, the actual mass which has crossed has to be divided by a molecular weight which is a weighted sum of the molecular weights of either the reactants or the products. This point is not important, however; what is helpful is to be able to regard $d\xi$ as the quantity of substance in molar terms which has crossed the chemical equation.

Now consider briefly the nature of equation (10.23). It can obviously be extended to the case in which there are any number of substances involved and not merely four. In words, it can be expressed thus: at chemical equilibrium the weighted sums of the chemical potentials of the reactants and of the products are equal, the 'weighting' being carried out by means of the stoichiometrical coefficients. In the transport case where equilibrium has not been attained, the quantity $(\mu_i^\alpha - \mu_i^\beta)$ can be looked upon as the 'driving force' behind the process; similarly in the chemical situation the quantity

$$A = (\nu_A\mu_A + \nu_B\mu_B - \nu_C\mu_C - \nu_D\mu_D) \text{ or } -\left(\frac{\partial G}{\partial \xi}\right)_{T,P},$$ often called the 'affinity'

of the reaction, can also be regarded as a driving force. At equilibrium the affinity is therefore zero (compare Figs. 10.3 and 10.4).

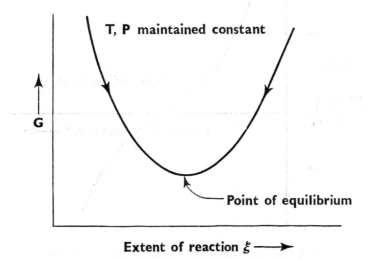

Fig. 10.3. Variation of the Gibbs free energy with the progress of a reaction.

Two further points remain. The conditions expressed in equations (10.24) and (10.25) for equilibrium are just as valid in referring to the end result of a complex, overall process as they are in a simple one. Provided genuine equilibrium holds, the equation

$$\mu_i^\alpha = \mu_i^\beta$$

will be true no matter how many phases and their associated membranes occur between α and β. Thus the cytoplasm may be a highly organized affair, replete with organelles and endoplasmic reticulum, and there may be added complexities of cell wall and tissue structure, but the condition for water equilibrium between vacuole and external milieu is quite unaffected, granted only that the metabolic processes involving energy release within the intervening region have no effect on water movement; that is, provided the requirement for genuine water *equilibrium* is met. And the same sort of thing holds of course for a series of chemical reactions.

The last comment is added to remove a possible misunderstanding. The equilibrium conditions expressed by equations (10.24) and (10.25) have nothing whatever to do with the system being maintained under conditions of constant temperature and pressure. They are valid under all conditions.

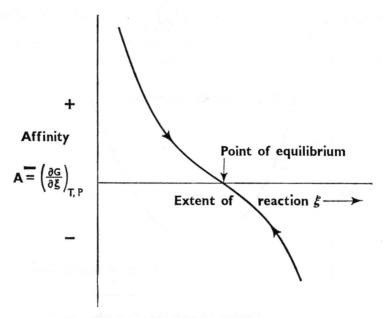

Fig. 10.4. Variation of the affinity with the progress of a reaction.

The idea of constancy of temperature and pressure only arose because we were considering, in effect, displacements from equilibrium and wished to use changes in the Gibbs free energy function (G) as a criterion. Other conditions (e.g. constant S, V) could equally well have been arranged for and then the alternative criteria employed for equilibrium. By virtue of the fact that μ is related to other quantities besides G (see equations (10.4), (10.7) and (10.8)) the final conclusion would have been just the same.

Having arrived at this point, the subject of the chemical potential will be left for development in the next chapter, when we shall consider how it can be expressed in terms of readily measurable quantities, and the use that can be made of it in interesting physiological situations.

CHAPTER 11

Chemical and Transport Processes in Dilute Solutions

> 'That's funny,' said Pooh. 'I dropped it on the other side,' said Pooh, 'and it came out on this side! I wonder if it would do it again?' And he went back for some more fir-cones.
>
> THE HOUSE AT POOH CORNER
>
> A. A. Milne

The properties of dilute solutions are of peculiar interest to both plant and animal physiologists. This arises from the facts firstly, that life is intimately bound up with reactions in the liquid phase, and secondly, that largely owing to the physico-chemical properties of the chief constituents of protoplasm, the proteins, the concentrations of the smaller chemical species are usually fairly low. It is true that there are cases where concentrations can be considerable—one thinks of the sugar in nectar and in sieve tube sap in higher plants, or of sodium chloride in salt marsh species—but broadly speaking the situation is as indicated. This means that the properties of dilute solutions have a prior claim on our attention.

DESCRIPTION OF COMPOSITION

The first thing that has to be done in approaching this subject is to decide on a method for describing the composition of a dilute solution. The most usual way of doing this in chemical and physiological work is to state the amount of solute, in grams or moles, which is present in unit volume (100 millilitres or 1 litre respectively) of solution. In the former case a percentage (w/v) composition is obtained; in the latter case, a molarity. Thus 5% (w/v) sodium chloride indicates a solution in which 5 grams of salt are present in 100 millilitres of solution; 0·5 M sucrose means that $0·5 \times 342 = 171$ grams of sucrose are present in 1 litre of solution. This method of specification has obvious advantages where volumetric analysis is being employed, or where solutions have to be progressively diluted to give a series of concentrations; but when theoretical considerations are involved it is highly inconvenient. For one thing, it still leaves in the air the question of how much solvent is present; it requires recourse to tables of density (which may not be available) and a tiresome calculation to ascertain this information.

Then again, volume is a quantity which is not conserved. It changes disconcertingly when systems are mixed, and of course with temperature and pressure also, so that a solution which is exactly molar at one temperature or pressure is not so at another. For these reasons, theoretical work always employs compositions specified on a basis of mass, and in this chapter we shall employ the molality m (as opposed to the mola*r*ity) defined as the mass of solute in moles associated with 1 kilogram of solvent (water). The reader will observe that for sufficiently dilute solutions the values of molality and molarity are very nearly the same, and to a first approximation they can usually be taken to be so. Further, it will be noticed that the molality of a solution with respect to a given solute is quite unaffected if a second solute is added. If the composition be specified by molalities m_1, m_2, m_3, ... then these are all independent variables, and there are as many of them as there are different solutes present. Finally, molality is an intensive property, independent of the size of the system.

Use of Moles Instead of Grams

The measurement of concentration in terms of moles rather than grams is suggested by the fact that the mole is related directly to the molecule, and molar ratios are therefore the same as molecular ones. This would clearly be expected, *a priori*, to lead to theoretical simplification, and this expectation is borne out by the phenomenon of colligative properties. These are properties which are found to depend to a first approximation on the *numerical* concentration of dissolved solute particles. Osmotic pressure, depression of the freezing point, and lowering of the vapour pressure are cases in point; they indicate that by deciding to work in molar terms we are probably making a wise choice, since molalities (or molarities) are an immediate indication of the numerical concentration of solute particles, whereas percentage compositions are not.

While on the subject it might be mentioned that probably the soundest simple measure of all is what is called the mole fraction (x). For a component i this is defined as the fraction whose numerator is the number (n_i) of moles (or molecules) of i present, and whose denominator is the number of moles (or molecules) of all sorts. In symbols,

$$x_i = \frac{n_i}{\Sigma n_i}. \tag{11.1}$$

That this is probably a sounder measure can be seen with reference to osmotic pressure. Bearing in mind that it is evidently the numbers of molecules present and not their nature that governs this property (that is, that it is colligative), we might enquire what happens to an osmotic pressure [1] if a

---

[1] It should be noted carefully that we are speaking about the actual osmotic *pressure*, not about the osmotic *potential*.

11. CHEMICAL AND TRANSPORT PROCESSES IN DILUTE SOLUTIONS

second solute is added, this solute being such that like the solvent, but unlike the first solute, it can freely traverse the membrane. Of course, this means that the second solute adds nothing to the osmotic pressure developed, and it is easy to see that its effect would be more or less comparable with that of adding an equal number of *solvent* molecules. In other words, a freely permeating solute takes its place alongside the solvent rather than alongside the non-permeating solute, in determining the osmotic pressure set up. This fairly obvious conclusion supports a preference for the mole fraction (x_i) rather than the molality (m_i), since the former changes when a second solute is added while the latter remains fixed at its old value.

However, the mole fractions are not quite so easy to handle as the molalities; in particular, when a new substance is added to a solution the mole fractions of all the old solutes promptly change, whereas their molalities do not. This is an important advantage for the latter, and when it is remembered that the discussion is limited to *dilute* solutions for which the advantages of the mole fraction are not very pronounced, it will be seen that the case for molalities is a strong one.

DEPENDENCE OF CHEMICAL POTENTIAL ON COMPOSITION

The next task is to find a formula connecting the chemical potential with such variables as the molality, temperature and pressure. If this can be done it will be possible to express the conditions of equilibrium found in the last chapter in more familiar terms than there employed, and so to make them correspondingly more directly useful.

If the reader casts his mind back to Chapter 5 he will recall that when a solution containing one mole of solute at a concentration[1] c^α is allowed to take in water so that its concentration falls to c^β, osmotic work can be extracted from the change (Fig. 6.1) whose maximum value under isothermal conditions is given by

$$W_{\text{max.}} = RT \ln \frac{c^\alpha}{c^\beta} \quad \text{(per mole).} \quad (11.2)$$

According to the definition of the chemical potential this expression must equal the change in the chemical potential ($\mu^\alpha - \mu^\beta$) of the solute between the two concentrations c^α and c^β. As a matter of fact, this conclusion is not absolutely obvious, although it may seem to be; and, as it is instructive to look into the matter a little more closely, we will do so.

The expression ($\mu^\alpha - \mu^\beta$) by definition represents the maximum work obtainable when one mole of solute is removed from a very large phase where

---

[1] Superscripts are employed rather than the previous subscripts since the latter will now be reserved for chemical species.

it exists at μ^α to a very large phase where it exists at μ^β, *there being movement of no other chemical species*. Now, in the harnessing of the process with an osmotic device there is a necessary movement of solvent, so that the condition would seem to be violated. What must be ensured is that after the use of the osmotic device there is *no net transfer of solvent* between the phases, only the solute having moved. This is a slightly more complicated problem, but it can be managed as follows.

Stage 1. We take a volume v^α of the first phase containing one mole of solute at c^α and allow it to absorb pure water (as in Fig. 6.1) till its concentration falls to c^β, the volume rising to v^β. We note that

$$v^\alpha c^\alpha = v^\beta c^\beta = 1,$$

and that since very dilute solutions are being dealt with the volume of water taken in is $(v^\beta - v^\alpha)$. The work gained is $RT \ln \dfrac{c^\alpha}{c^\beta}$.

Stage 2. The sample at concentration c^β is now simply added to the second phase; this would complete the transfer were it not for the fact that solvent has also been transferred, and a quantity of initially pure water has been used up. In these respects the *status quo* has to be restored, and this is done in two further stages.

Stage 3. We apply its own osmotic pressure (equal to RTc^β) to phase β confined behind a semi-permeable barrier. This will enable a volume v^β of pure water to be 'squeezed out' reversibly, the work required being $RTc^\beta \cdot v^\beta = RT$.

Stage 4. We take a volume v^α of this water (leaving the correct amount pure) and add this reversibly to phase α. The work to be gained from this addition is similarly $RTc^\alpha \cdot v^\alpha = RT$, and this exactly balances the amount involved in *Stage* 3.

The necessary *status quo* has now been completely restored, and it follows that it is justifiable to write

$$\mu^\alpha - \mu^\beta = RT \ln \frac{c^\alpha}{c^\beta}, \tag{11.3}$$

or what amounts to the same thing for a very dilute solution

$$\mu^\alpha - \mu^\beta = RT \ln \frac{m^\alpha}{m^\beta}, \tag{11.4}$$

where c^α, c^β are molarities, and m^α, m^β molalities.

There are some important things to notice about formulae (11.3) and (11.4). In the first place, in deriving them no mention has been made of the nature of either the solvent or the solute; consequently these formulae have a very general validity as descriptive of the dependence of chemical potential on composition. Further, they are accurate in so far as the expression assumed for the osmotic pressure of a solution (i.e. $\Pi = RTc$ or RTm) is accurate. The reader will be aware that this relationship for many solutions becomes inexact fairly rapidly as the concentration rises, so clearly expressions (11.3) and (11.4) will normally only be approximately true. The question naturally arises as to which of them is the better approximation, or whether there is any other which would be better still. The answer is that the best simple approximation available to us is obtained by using x_i, the mole fraction of the species i, rather than m_i, the molality; c_i is undesirable for the reasons given earlier. That x_i is better than m_i can be appreciated by the sort of arguments used before; and this leads to the subject of what are called ideal solutions. These embody in the realm of solutions some of the properties of ideal gases, and in line with what has been said they are defined by inserting x instead of m into equation (11.4) and then regarding this equation as exact. In other words, an ideal solution is one for which the chemical potential of the species i is given *exactly* by the equation

$$\mu_i - \mu_i^0 = RT \ln x_i \qquad (11.5)$$

where one of the two phases is taken for convenience as the pure substance (corresponding to $x_i = 1$) and indicated by the superscript zero. No real solution is exactly ideal, but solutions or mixtures [1] in which all the molecules present (both solute and solvent) are chemically similar may come very close to ideality, even at high concentrations. They therefore provide a very useful basis of comparison with real solutions.

In defining an ideal solution one of the phases α, β was made a 'standard' one, with x_i equal to unity. The corresponding chemical potential then takes a 'standard' value, and this serves to fix the arbitrary constant to which, as mentioned in the previous chapter, the chemical potential (as well as U, S, F and G) is subject. Henceforward molalities (m_i) will be dealt with instead of mole fractions (x_i), and it is convenient to define a new standard with m_i equal to unity. This is quite different from the previous standard since $x_i = 1$ indicates the pure substance, whereas $m_i = 1$ is approximately a molar solution. If the symbol μ_i^0 is kept for the pure substance another is needed for its potential in molal solution, and the symbol $\mu_i^\ominus$ ('mu-i-plimsoll') is often used, suggested by the Plimsoll or standard loading line on merchant ships. With this notation, the equation

$$\mu_i = \mu_i^\ominus + RT \ln m_i \qquad (11.6)$$

may be written, and the reader can remind himself of the meaning of

---

[1] A solution is merely a mixture in which one component (the solvent) is understood to be present in relative excess.

$\mu_i^\ominus$ by merely writing $m_i = 1$ in this equation, when the last term vanishes and we have $\mu_i = \mu_i^\ominus$. Needless to say $\mu_i^\ominus$ is only a constant so far as change of composition is concerned; it will vary with changes of pressure and of temperature. Especially in the case of pressure this is something the physiologist must remember.

IDEAL DILUTE SOLUTIONS

Solutions for which equation (11.6) is *exact* over the lower range are often called 'ideal dilute'. They cannot be ideal in the unqualified sense already described since, unlike the mole fraction x_i which can vary only from zero to one, the molality m_i has a range from zero to infinity (the value infinity corresponding to the pure substance). Since the chemical potential of a pure substance can hardly be infinite (because a virtually pure substance may exist in equilibrium with its saturated solution, in which its potential is naturally regarded as finite), it can be seen that such an equation as (11.6) cannot fit the physical facts over the entire range of values which m_i takes. Thus, the most that can be done is to define a class of solutions which obey equation (11.6) exactly up to the (low) concentration of interest, and such solutions are called ideal dilute. Needless to say, real solutions never quite fit into this category, but it will often be supposed for simplicity that they do. Corrections can be applied in actual cases if necessary.

A HYDROSTATIC ANALOGY

Professor Everett in his excellent book, (1959)[1], has a useful hydrostatic analogy for equation (11.6). Imagine a vertical exponential horn (Fig. 11.1) the form reflecting the logarithmic relationship [2] between μ and m. Such a horn extends downwards in the narrowing direction to infinity. If liquid is poured into the empty horn its level at first rises extremely rapidly. This corresponds to a very rapid rise in the chemical potential when solute is first added to pure solvent; with successive additions the level in the horn and the chemical potential in the solution rise less and less steeply. If there are two horns representing two phases, the liquid heights standing at h_1 and h_2 respectively, the transfer of a small mass dm of liquid from one horn to the other represents work [3] given by

$$dW = dmg(h_1 - h_2) \qquad (11.7)$$

[1] Everett, D. H. (1959). 'An Introduction to the Study of Chemical Thermodynamics.' Longmans, London.

[2] Equation (11.6) can also be written in exponential form: $m_i = \exp\left(\dfrac{\mu_i - \mu_i^\ominus}{RT}\right)$.

[3] The weight, i.e. gravitational force, of a mass m is mg, where g is the acceleration due to gravity.

since the liquid can be regarded as being skimmed off one surface and added to the other a distance (h_1-h_2) below it. This corresponds to the previous formula:

$$\text{Work} = dn_i(\mu_i^\alpha - \mu_i^\beta) \tag{11.8}$$

where dn_i moles are transferred from one chemical potential to another. There is therefore a parallelism between h in the model and μ_i in the solution. Corresponding to $\mu_i^\ominus$ there will be an arbitrary level $h^\ominus$ (Fig. 11.1) which provides a base line (like 'sea level') above which to measure heights. Further, the volume of liquid in the horn is analogous to the concentration of the solute i, so that the volume of the horns up to the level $h^\ominus$ must be unity.

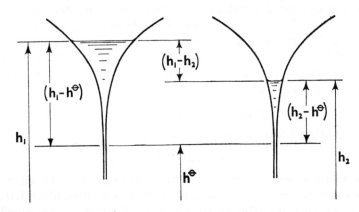

FIG. 11.1. A hydrostatic analogy for the chemical potential of a solute.

DIFFERENCE OF CHEMICAL POTENTIALS

It need hardly be stressed that the difference [1] between two chemical potentials only has a *physical* meaning under the following circumstances:

(i) they must both refer to the same substance, or to different substances which can be converted chemically one into the other. (In the latter case we can have combinations of potentials such as $(\nu_A\mu_A + \nu_B\mu_B - \nu_C\mu_C - \nu_D\mu_D)$.) This means that actual physical matter must be capable of passing from the condition corresponding to one potential into the condition corresponding to the other. Such an expression as $(\mu_\text{water} - \mu_\text{sulphur})$ is *physically meaningless*[2] for this reason;

(ii) they must both refer to the same *temperature*, but they need not refer to the same *pressure*. The distinction arises here from the fact that any barrier which will allow matter to pass will also of necessity allow heat to pass, but the parallel statement is not necessarily true. If an attempt is made to

---

[1] Differences are discussed because single chemical potentials are subject to an unknown additive constant.

[2] For remarks on 'physically meaningless' see below (p. 156).

evaluate the maximum work $dn(\mu^\alpha - \mu^\beta)$ associated with the movement of dn moles from μ^α to μ^β, where a temperature difference is present it is found that an undefined amount of heat [1] is also involved and this renders the calculation quite indefinite; in fact there *is* no maximum amount of work. This difficulty clearly does not arise with pressure differences;

(iii) where the two potentials refer to a solute they must both (normally) be associated with the same solvent (we are still confining our attention to solutions and forgetting the more general category of mixtures). That is, we cannot simply say (on the grounds of equation (11.3)) that the difference in chemical potential of solute between two solutions in *different* solvents is $RT \ln$ (ratio of concentrations), as can be done if they are in the same solvent. However, misunderstandings must be avoided at this point; there *is* a physical meaning in the difference in chemical potentials of solute in two different solvents at the same temperature, but the situation is not simply described by equation (11.3) or (11.4). The reason for this is as follows. For one solution α, (say urea in water), from (11.6)

$$\mu^\alpha_{\text{urea}} = \mu^\ominus_{\text{urea}} + RT \ln m^\alpha_{\text{urea}}$$

and for the other β, (say urea in olive oil),

$$\mu^\beta_{\text{urea}} = \mu^\ominus_{\text{urea}} + RT \ln m^\beta_{\text{urea}}.$$

If both solutions were aqueous these two equations could be subtracted and the constant terms ($\mu^\ominus$) would cancel, giving a relation like (11.4); but since the solvents are different *the two standard values are not equal*[2]. In other words a molal solution of urea in water would not be in equilibrium with a molal solution (if one were possible) in olive oil. If they were shaken up together urea would pass from the oil to the water, and a partition experiment of this sort would have to be done to determine how to relate the two standard chemical potentials to each other. It would not be a difficult matter, but fortunately the physiologist has to deal almost exclusively with a single solvent, water, so the problem does not arise very often.

Dependence of Chemical Potential on Temperature and Pressure

Bearing in mind that the chemical potential of a component measures the 'urge' of that component to escape from where it is (either by a transport process or through a 'chemical back door') it is understandable that the chemical potential of a species rises with an increase in either the temperature

---

[1] The actual amount of heat involved depends on which one of an infinite number of possible experimental sequences is followed.

[2] They should really therefore have different symbols.

or the pressure. What this means is that if two phases (α and β, Fig. 11.2) are in equilibrium with one another with respect to a substance i, then if either the pressure or the temperature of one phase (say α) is raised (to a steady value) while that of the other is left unaltered the system will be thrown out of balance and there will be a movement of i from α to β. In the case of a steady imposed temperature rise this movement can never lead to a state of equilibrium for the reason mentioned earlier, that it is impossible to envisage arrangements whereby matter can be transported without heat being allowed to pass as well. Consequently the temperature difference will either disappear

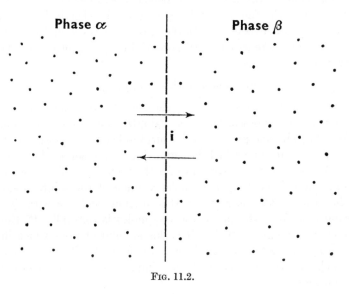

Fig. 11.2.

by conduction, or it will necessitate a continual supply (and removal) of heat to sustain it, and this of course is the very denial of equilibrium. This situation is quite an important one, but it requires to be dealt with by the newer development called the Thermodynamics of Irreversible Processes, and this will be left till the last chapter.

The situation is quite different, however, in the case of a pressure increase. Owing to the possibility of achieving semi-permeable membranes, true equilibrium can result. What happens [1] is that the pressure rise in phase α 'squeezes' some of component i into phase β, and the movement comes to an end in true equilibrium when the increase of μ_i^α due to the rise of pressure in phase α is balanced by its decrease due to the fall in concentration. (It can be assumed for simplicity that phase β is very large and so remains

[1] It can be supposed for simplicity that the membrane separating phases α and β is permeable only to component i.

unchanged.) This is, of course, just what happens when osmotic pressure is developed, but for the moment solutes are being considered rather than solvent, which is the component transported in osmosis.

It is quite a simple matter to derive a formula to show how the chemical potential varies with the pressure. This means evaluating $\left(\dfrac{\partial \mu_i}{\partial P}\right)_{T,n_i,n_s}$ where the subscripts n_i, n_s mean that the contents of the solute i and the solvent s are held constant (for simplicity it is assumed that there is only one solute). The value of this differential coefficient can be derived in two ways. Firstly, recalling the fundamental property of the chemical potential that the maximum work associated with the movement of dn_i moles of substance from a place (α) where its chemical potential is μ_i^α to a place (β) where it is μ_i^β is given by

$$\text{maximum work} = dn_i(\mu_i^\alpha - \mu_i^\beta)$$

it will appear that if the pressure is raised at the first place by an amount dP^α, an extra amount of work equal to $\overline{V}_i dn_i dP^\alpha$ is involved in escaping from α. To see this, imagine phase α to be surrounded by an elastic wall, which applies the pressure and which is permeable to the solute, but not to the solvent. When dn_i moles moves out of α the volume of this phase decreases by $\overline{V}_i dn_i$, where $\overline{V}_i$ is the partial molar volume of the solute (Chapter 3). Thus the elastic wall contracts by this amount, and the work it does in 'squeezing out' from the phase the dn_i moles is naturally P^α times this volume contraction, i.e. $P^\alpha \overline{V}_i dn_i$. It is this amount of work which increases by $dP^\alpha \overline{V}_i dn_i$ when P^α is increased by an amount dP^α. With phase β undergoing no change, the increase in the maximum work can be expressed in the equivalent ways

$$\text{Increase} = dn_i d\mu_i^\alpha = \overline{V}_i dn_i dP^\alpha$$

from which it is easy to obtain at once, by dropping the α as no longer necessary and writing in the partial differential form for explicitness,

$$\left(\dfrac{\partial \mu_i}{\partial P}\right)_{T,n_i,n_s} = \overline{V}_i. \tag{11.9}$$

Thus, it all depends on the effect that addition of the solute has on the total volume; the chemical potential of solutes with large molecules will normally be more sensitive to pressure than that of solutes with small ones.

Equation (11.9) can be derived more rigorously, but less instructively, as follows, using the cross-differentiation identity (equation (3.24)). By definition,

$$\mu_i = \left(\dfrac{\partial G}{\partial n_i}\right)_{T,P,n_s}. \tag{11.10}$$

Differentiating with respect to P at constant T, n_i, n_s it is found that

$$\frac{\partial \mu_i}{\partial P} = \frac{\partial}{\partial P}\left(\frac{\partial G}{\partial n_i}\right)$$

$$= \frac{\partial}{\partial n_i}\left(\frac{\partial G}{\partial P}\right) \tag{11.11}$$

using the identity mentioned. However, for a closed phase in equilibrium (the phase can be considered closed because $\dfrac{\partial G}{\partial P}$ is being evaluated at constant content) there is the fundamental relation (equation (8.15))

$$dG = VdP - SdT \tag{11.12}$$

and if dT is made zero (T being constant) this gives

$$\left(\frac{\partial G}{\partial P}\right)_{T,n_i,n_s} = V. \tag{11.13}$$

Inserting this in equation (11.11) it follows at once that

$$\left(\frac{\partial \mu_i}{\partial P}\right)_{T,n_i,n_s} = \left(\frac{\partial V}{\partial n_i}\right)_{T,n_s} = \overline{V}_i \tag{11.14}$$

as before.

Case of Turgor Pressure

A physiological situation in which the foregoing considerations may occasionally be important arises in connection with turgor pressure. The pressure of the cell sap can sometimes reach fairly high values, and quite apart from anything else this will influence the distribution of a solute between the sap and the external medium. To see the magnitude of this effect imagine a tissue immersed in a solute which is inert but able to penetrate readily into the vacuoles. If this solute has a molar volume about ten times that of water (corresponding to a simple sugar), then $\overline{V}_i$ can be taken as approximately 180 millilitres, or say 0·2 litre. What has to be evaluated is, in effect, $\left(\dfrac{\partial m_i}{\partial P}\right)_{\mu_i}$; that is, by how much the (internal) molality (m_i) must be altered to balance the effect of the turgor pressure [1] (dP) under conditions in which the chemical potential remains fixed at the value of the external medium. The non-turgid cells can be thought of as initially having a vacuolar concentration of solute

---

[1] The turgor pressure is not the total pressure P, but the pressure above ambient, and so here corresponds to dP.

i equal to that of the medium, and being in equilibrium in respect of i. Water then enters and the vacuolar pressure rises by an amount dP. To maintain μ_i inside equal to the external value, solute leaves the vacuole, the concentration changing by dm_i. Thus the ratio $\dfrac{dm_i}{dP}$ is really $\left(\dfrac{\partial m_i}{\partial P}\right)_{\mu_i}$, as suggested above.

Now since μ_i is a function of P and m_i (T is of course not being considered as a variable), we have (see equations (3.20) and (3.21))

$$\left(\frac{\partial m_i}{\partial P}\right)_{\mu_i} = -\left(\frac{\partial \mu_i}{\partial P}\right)_{m_i} \Big/ \left(\frac{\partial \mu_i}{\partial m_i}\right)_P. \tag{11.15}$$

But $\left(\dfrac{\partial \mu_i}{\partial P}\right)_{m_i}$ as we have just seen is equal to $\overline{V}_i$. Further, from equation (11.6)

$$\left(\frac{\partial \mu}{\partial m_i}\right)_P = \left(\frac{RT}{m_i}\right) \tag{11.16}$$

since with T and P constant $\mu_i^\ominus$ is a constant. Thus substituting in equation (11.15) we find

$$\left(\frac{\partial m_i}{\partial P}\right)_{\mu_i} = -\frac{m_i \overline{V}_i}{RT}$$

or

$$\frac{1}{m_i}\left(\frac{\partial m_i}{\partial P}\right)_{\mu_i} = -\frac{\overline{V}_i}{RT}. \tag{11.17}$$

The left hand side of this equation represents the *proportional* change in molality per unit of turgor pressure, and in the case under consideration, putting $\overline{V}_i = 0.2$ litre, $R = 0.082$ litre atmospheres per mole per °C, and $T = 293°K$ this becomes

$$-\frac{0.2}{0.082 \times 293} = -0.0083 \text{ or } -0.8\% \text{ per atmosphere.}$$

Thus with a turgor pressure of 10 atmospheres the equilibrium concentration inside would be about 8% less than that outside. This applies of course to a *penetrating* solute whose molecules are of the size mentioned.

Remarks on 'Physically Meaningless'

Before passing on to the next subject there is a point that must be considered. The reader may have been puzzled by the earlier statement that a difference in two chemical potentials has no physical meaning if the two potentials refer to the same substance at different temperatures, or to two quite unrelated substances at the same temperature. This statement may give rise to some misunderstanding, since it is implied in the general equation for an open phase (see equation (10.15)):

$$dG = VdP - SdT + \mu_1 d\mu_1 + \mu_2 dn_2 + \ldots \qquad (11.18)$$

that it is permissible to add (and of course subtract) potentials for quite unrelated substances[1]. Further, in certain formulae differences in chemical potentials appear in which these potentials relate to different temperatures. Obviously then such expressions as $(\mu_{\text{water}} - \mu_{\text{sulphur}})$ or $\{\mu_i(T_1) - \mu_i(T_2)\}$ cannot be meaningless. What is meant by saying that they are *physically* meaningless is that no formula in which they appear can be submitted to direct physical verification; it can only occur as an intermediate stage in the theory, and by the time a result has emerged capable of being tested quantitatively by laboratory measurements such expressions have invariably disappeared. All this can be seen very clearly with reference to the temperature dependence of the chemical potential. If the same method is followed as in the preceding paragraph, then starting from the fundamental equation (11.12)

$$dG = VdP - SdT \qquad (11.19)$$

but this time making dP equal to zero, we find

$$\left(\frac{\partial G}{\partial T}\right)_P = -S. \qquad (11.20)$$

This should be compared with equation (11.13). Using the cross differentiation identity gives

$$\frac{\partial \mu_i}{\partial T} = \frac{\partial}{\partial T}\left(\frac{\partial G}{\partial n_i}\right)$$

$$= \frac{\partial}{\partial n_i}\left(\frac{\partial G}{\partial T}\right)$$

$$= -\frac{\partial S}{\partial n_i} = -\overline{S}_i \qquad (11.21)$$

introducing the result (11.20). The quantity $\overline{S}_i$ is the partial molar entropy of component i (see Chapter 3). Now the derivative $\dfrac{\partial \mu_i}{\partial T}$ involves comparing values of μ_i at different temperatures, which is one of the things we have been discussing. Further, the partial molar entropy $\overline{S}_i$ to which it is related by equation (11.21) is unknown to the extent of an additive constant. As a result it is also impossible to attribute a numerical value (except a quite

---

[1] If dP and dT are made zero, and dn_1 and dn_2 are made equal and opposite, the quantity $(\mu_1 - \mu_2)$ appears at once and so must be meaningful if the equation is to be so.

arbitrary one) to $\dfrac{\partial \mu_i}{\partial T}$, and this implies that neither can an absolute numerical value be attributed to such an expression as $\{\mu_i(T_1)-\mu_i(T_2)\}$. If it were possible to evaluate this last expression unequivocally from a laboratory experiment, there would be means at our disposal for finding the absolute value of the entropy of a system, which as was seen earlier is quite impossible. This is what was meant by saying that such differences of chemical potential as $(\mu_{\text{water}}-\mu_{\text{sulphur}})$, or $\{\mu_i(T_1)-\mu_i(T_2)\}$ are physically meaningless.

The Free Energy of Chemical Reactions

As every physiologist knows the energy, or rather the free energy, which maintains the life of the cell comes predominantly from chemical reactions within it. He is familiar with the idea that energy from respiration is employed in the synthesis of protein or reserve polysaccharides, in the accumulation of salts, and probably in the translocation of assimilates; thus chemical energy is linked with the doing of what may be called chemical work, osmotic work and frictional work. Of course, chemical free energy is not the only source of energy relevant to the life of the plant. Photosynthesis utilizes the energy of radiation; transpiration (and so much of the upward movement of minerals) depends on the free energy of evaporation; and volume expansion of tissues utilizes osmotic free energy. However, it is probably not too much to say that chemical free energy occupies quite a pre-eminent position in the work economy of living systems. Thus the subject of the free energy changes associated with chemical reactions is an important one, for it must not be forgotten that it embraces the *storage* of energy as well as its *release*, and thus claims the major stake even in photosynthesis. The reader need hardly be reminded in all this that it is the Gibbs free energy (G) which is being referred to almost exclusively because, in the physiological context, processes nearly always take place under conditions which approximate very closely indeed to constancy of temperature and pressure; and because under these conditions the *sole determining factor* with regard to the spontaneity of a process is whether or not it lowers G. No other considerations enter; a knowledge of how G varies gives everything that it is necessary to know about the direction in which the system will tend [1] to move.

The Specification of the Free Energy Change

The first problem that we face is how to specify a chemical change in order to be able to attribute to it an unequivocal value of the free energy

[1] Whether of course it *does* move in this direction, and at what rate, will depend *inter alia* on whether appropriate enzymes are present, or membrane barriers absent.

11. CHEMICAL AND TRANSPORT PROCESSES IN DILUTE SOLUTIONS

involved. Something more is required than merely to state what the reaction is, for obviously if a chemical system happens to be at its equilibrium point the chemical reaction can be induced to move infinitesimally either one way or the other without involving any free energy change at all. Clearly the concentrations must also be stated, or alternatively the chemical potentials, at which the various substances are supposed to react.

Consider the simple reaction used before:

$$\nu_A A + \nu_B B \rightarrow \nu_C C + \nu_D D \qquad (11.22)$$

where the ν's are the stoichiometrical coefficients. Suppose either, that the system is so large that the actual amounts of the substances stated in the equation (i.e. ν_A moles of A, and so on) can react from left to right without altering any of the concentrations by significant amounts; or alternatively, that it is of normal size but that minute amounts $\nu_A d\xi$, $\nu_B d\xi$ etc., react, where $d\xi$ is infinitesimally small. With either of these suppositions the chemical potentials will remain steady; and (choosing the larger system for simplicity) the change in free energy when the amounts ν_A, ν_B cross to the other side of the equation will be given by

$$\Delta G = -\nu_A \mu_A - \nu_B \mu_B + \nu_C \mu_C + \nu_D \mu_D. \qquad (11.23)$$

This follows from the definition of μ_i as $\dfrac{\partial G}{\partial n_i}$. Now replace the μ's by the expressions (11.6) previously derived for them in terms of molality:

$$\mu_A = \mu_A^\ominus + RT \ln m_A \qquad (11.24)$$

and so on. This leads to

$$\Delta G = -\nu_A \mu_A^\ominus - \nu_B \mu_B^\ominus + \nu_C \mu_C^\ominus + \nu_D \mu_D^\ominus + RT \ln \left(\frac{m_C^{\nu_C} m_D^{\nu_D}}{m_A^{\nu_A} m_B^{\nu_B}} \right). \qquad (11.25)$$

This is a general equation holding for all values of the m's. It shows by how much the free energy of the system changes for a given amount of reaction (ν_A moles of A, etc.) when the reaction takes place at different concentration levels. However it contains the unknown[1] standard values $\mu_A^\ominus, \mu_B^\ominus, \mu_C^\ominus, \mu_D^\ominus$ and it would be convenient if these could be eliminated in favour of some constant or constants for the reaction, accessible to physical measurement. Fortunately this can be done very easily. Suppose any set of values of the m's is inserted in this equation which satisfy the equilibrium condition

$$\frac{m_C^{\nu_C} m_D^{\nu_D}}{m_A^{\nu_A} m_B^{\nu_B}} = K \qquad (11.26)$$

[1] In absolute magnitude.

where K is the equilibrium constant[1]. For this particular set of values ΔG is naturally zero. Thus introducing (11.26) into (11.25) we have

$$0 = -\nu_A\mu_A^\ominus - \nu_B\mu_B^\ominus + \nu_C\mu_C^\ominus + \nu_D\mu_D^\ominus + RT\ln K. \tag{11.27}$$

If equation (11.25) is subtracted from this the objectionable standard potentials disappear, and instead we have

$$-\Delta G = RT\ln K - RT\ln\left(\frac{m_C^{\nu_C}m_D^{\nu_D}}{m_A^{\nu_A}m_B^{\nu_B}}\right) \tag{11.28}$$

where the negative value $-\Delta G$ has been given since interest is usually in the free energy *decrease*.

Several important points emerge from this equation[2]. Firstly, the form in which the m's enter is exactly the same as that in which they enter into K, and it follows from this as a practical consequence that if molalities are to be used in this equation rather than volume concentrations then K must be calculated from data expressed also in molalities; and similarly if volume concentrations (or for gases, partial pressures) are to be used.

Secondly, it is worth remarking that equation (11.28) is really a proof, by thermodynamic arguments, of Guldberg and Waage's famous law of mass action. For if the m's happen to represent a different set of *equilibrium* values from those used to calculate K, then since the system is at equilibrium ΔG is zero and so

$$RT\ln K - RT\ln\left(\frac{m_C^{\nu_C}m_D^{\nu_D}}{m_A^{\nu_A}m_B^{\nu_B}}\right) = 0$$

or

$$\frac{m_C^{\nu_C}m_D^{\nu_D}}{m_A^{\nu_A}m_B^{\nu_B}} = K. \tag{11.29}$$

This indicates that *any* set of values of the m's which results in chemical equilibrium will yield the same value for K as the set actually used to calculate it; in other words, K is a definite constant relating to the particular equilibrium. This result, which can easily be derived by statistical arguments based on molecular theory, follows thermodynamically as has been seen from considerations of the connection between equilibrium and free energy; consequently one is led to suspect that thermodynamic properties are bound up somewhere with the statistical relations of molecules. In the particular case of entropy, this point has been discussed at length (Chapter 9).

Thirdly, the important point emerges that if the molalities of all the reactants (and products) happen to be unity in a very large system, then the

[1] This is an equation of chemical kinetics also and its derivation can be found in any good elementary textbook.

[2] The equation is one form of the van't Hoff isotherm.

free energy change which occurs when the amounts specified by the equation (ν_A moles of A, etc.) pass into products is given directly by

$$-\Delta G^\ominus = RT \ln K. \qquad (11.30)$$

Bearing in mind that the standard potential, $\mu_i^\ominus$ was defined as that belonging to *unit* molality, the reader will not be surprised at the symbolism ($-\Delta G^\ominus$) and the name (standard free energy change) which is employed here. There are, to complicate matters, as many different values for the standard free energy as there are different ways of specifying the 'concentration' of the reactants. The value will be different according as these are measured in mole fractions, molalities, partial pressures or concentrations per unit volume. The point is not important, as it corresponds merely to measuring heights above a different base level; but it does mean that one must be consistent. To assist in this authors commonly use different symbols (ΔG^0, $\Delta G^\ominus$, $\Delta G\dagger$ are some) to indicate the different standards, and here we have adopted the plimsoll superscript to connote the use of molalities.

The reader can easily verify from equations (11.28) or (11.30) that where the equilibrium constant is extremely large (this can be regarded as the statistical way of saying that the equilibrium lies very far over to the right), the free energy *decrease* is also extremely large (this is the thermodynamic way of saying the same thing). Oppositely, a very minute equilibrium constant will mean a large free energy *increase*.

Value of the Stoichiometrical Coefficients

Since the value of the free energy change refers to the actual transformation of ν_A moles of A and ν_B moles of B it must be taken in conjunction with the actual values of these coefficients. Thus, if the whole chemical equation were to be multiplied through by two it would still 'balance' and represent (apparently) the same reaction; but the free energy change corresponding to it would be doubled. This is not inconsistent with equation (11.30), since the reader can easily verify that the value of K has been squared, and $\ln K$ is therefore also doubled. A case in point where a mistake might be made is in the free energy of formation of water where the reaction is sometimes written

$$H_2 + \tfrac{1}{2}O_2 \to H_2O$$

and sometimes

$$2H_2 + O_2 \to 2H_2O.$$

There is no real difficulty here, but care must obviously be exercised to see that the values of the coefficients taken, and the free energy change assumed, do correspond. As a matter of fact when a chemical equation has been multiplied through by an integer it does not represent quite the same reaction as before, since its 'order' has been changed. But this is a matter for chemical kinetics to pronounce upon, and so it lies outside the scope of our subject.

RELATION OF THE AFFINITY TO THE FREE ENERGY CHANGE

In the previous chapter the affinity[1], A of a chemical reaction was mentioned where

$$A = -\left(\frac{\partial G}{\partial \xi}\right)_{T,P} = \nu_A\mu_A + \nu_B\mu_B - \nu_C\mu_C - \nu_D\mu_D \qquad (11.31)$$

in which $d\xi$ may be looked upon as the number of moles of substance passing from left to right of the chemical equation calculated on a molecular weight of $(\nu_A M_A + \nu_B M_B)$. The affinity can be regarded as the 'urge' behind the reaction, and the standard affinity, $A^\ominus$ is its value when all reactants and products happen to be at unit molality. The reader may wonder what is the relation between $A^\ominus$ and $-\Delta G^\ominus$ (or A and $-\Delta G$). The answer is that they are numerically exactly equal, but represent rather different ways of looking at the same situation. When $-\Delta G$ is used the system is thought of as large enough to allow ν_A, ν_B moles of A and B to react without altering it qualitatively (so that all the μ's may be regarded as steady). When the affinity A is used the ordinary small system is being considered but it is imagined that only $\nu_A d\xi$, $\nu_B d\xi$ moles to react, these infinitesimal amounts being naturally too minute to affect the μ's. Then the infinitesimal dG is taken and 'proportioned up', as it were, by dividing by $d\xi$ to give a value equal to ΔG; this corresponds to $\Delta \xi = 1$. The affinity is thus a rather better conception mathematically than the 'free energy change', but numerically the two are exactly the same.

ADDITION OF THERMOCHEMICAL EQUATIONS

What the physiologist or biochemist almost always requires to know is the affinity (or the free energy change[2]) of a particular reaction under the conditions *actually existing*. He has therefore to ascertain the *standard* affinity from published data (and these will often be available only for reactions other than his own) and then to adjust it to allow for the fact that actual molalities are not all unity. In this programme he has usually at some stage to invoke a fundamental property of equations expressing thermodynamic data: they can be added and subtracted according to very simple rules, and

---

[1] The affinity is also equal to $-\left(\frac{\partial F}{\partial \xi}\right)_{T,V}$, $-\left(\frac{\partial U}{\partial \xi}\right)_{S,V}$ and $-\left(\frac{\partial H}{\partial \xi}\right)_{S,P}$.

[2] The term 'affinity' will frequently be used henceforth, but the reader can substitute the more usual 'free energy change' if he wishes.

this applies not only to chemical equations, but also to all types of physical and physico-chemical ones. In connection with what are often called thermochemical data (heats of combustion), the possibility of doing this was first realized by Hess and is often called Hess's law. This states that 'the heat change in a chemical reaction is the same whether it takes place in one or several stages'. Owing to the fact that it was first stated for heat energy it is often thought that the law applies only because the property in question is subject to conservation, as energy is. Actually it applies to all extensive properties which are *properties of state*, whether they are subject to conservation or not. The statement could therefore be enlarged to read as follows: 'the change in energy (or volume or entropy or free energy) in a chemical (or physico-chemical or physical) reaction is the same whether it takes place in one or several (practicable or impracticable) stages'. In fact, it may sometimes help the reader in manipulating the addition and subtraction of thermodynamic equations to imagine for the moment that they relate to something familiar, like volume changes, rather than to free energy.

The rules which were referred to are simply these. Firstly, the change in the thermodynamic property is set alongside the chemical equation with its sign correct (this must correspond with the reactants being on the left, and the products on the right). One or more other equations are then set below the first and arranged similarly. The equations are added or subtracted according to the usual laws of algebra; terms may be cancelled when they occur on both sides of the final equation but only when they represent matter identical in quality and quantity, and this may necessitate multiplying some of the equations first by numbers. The rules are really quite commonsense, but what has been just said about quality needs emphasizing.

Suppose for instance it was necessary to ascertain what the evolution of heat would be if a catalyst could be found to promote the (quite impossible) reaction[1]:

$$2C + 3H_2 + \tfrac{1}{2}O_2 \rightarrow C_2H_5OH. \tag{11.32}$$

By heat here is meant the amount of energy evolved when the mixture is made to react in a calorimeter open to the atmosphere. The reader will recall that under these conditions (constant pressure and substantially constant temperature) the calorimeter measures $-\Delta H$. Since the direct reaction is impracticable the process must be carried out in indirect stages. Three reactions easy to arrange are

(i) $C_2H_5OH(l) + 3O_2(g) \rightarrow 2CO_2(g) + 3H_2O(l)$; $\Delta H = -327.0$ kcal

(ii) $C(s) + O_2(g) \rightarrow CO_2(g)$; $\Delta H = -94.0$ kcal

(iii) $H_2(g) + \tfrac{1}{2}O_2(g) \rightarrow H_2O(g)$; $\Delta H = -57.8$ kcal.

[1] Glasstone, S. (1947). 'Textbook of Physical Chemistry', p. 205. D. Van Nostrand Co., New York; Macmillan, London.

It will be noticed that details of whether the substances are solid (s), liquid (l) of gaseous (g) have been inserted, since this will obviously reflect on the values of ΔH; the figure alongside equation (i) would be about 337·0 instead of 327·0 if the alcohol were given its latent heat first and allowed to react as a gas. On the other hand, it makes very little difference at what precise pressure the reactants are introduced, so this need not be specified. Equation (11.32) can (almost) be arrived at by multiplying (i) by minus one (that is, reversing it), (ii) by two, and (iii) by three, and adding. The result is, after cancelling identical terms on opposite sides,

(iv) $2C(s) + 3H_2(g) + \frac{1}{2}O_2(g) + 3H_2O(l) \rightarrow C_2H_5OH(l) + 3H_2O(g)$;
$\Delta H = +327\cdot0 + 2(-94\cdot0) + 3(-57\cdot8) = -34\cdot4$ kcal.

It should be noted that the terms $3H_2O(l)$ and $3H_2O(g)$ cannot be cancelled since the conversion of one into the other is associated with a considerable ΔH, in fact with the latent heat of three moles of water (31·5 kcal). Therefore it is necessary to introduce a further equation (incidentally not a chemical reaction at all):

(v) $3H_2O(l) \rightarrow 3H_2O(g)$; $\Delta H = 31\cdot5$ kcal

which subtracted from (iv) finally gives

(vi) $2C(s) + 3H_2(g) + \frac{1}{2}O_2(g) \rightarrow C_2H_5OH(l)$; $\Delta H = -65\cdot9$ kcal.

Thus the evolution of heat which would occur if its solid and gaseous elements could be directly combined in an open calorimeter to give liquid alcohol is 65·9 kcal per mole of product; and if conversely it could be split up into its elements the *absorption* of heat would be the same.

Application to Free Energy

Exactly the same procedure as the previous one is applicable to free energy calculations. However, there are minor differences. Biological reactions are almost always reactions in aqueous solution. Now when a common substance like sucrose is dissolved in water there is very little change in temperature; in other words for the reaction[1]

$$\text{Sucrose} + \text{water} \rightarrow \text{sucrose, aq.} \qquad (11.33)$$

ΔH is negligible. Consequently it would be justifiable to cancel '$C_{12}H_{22}O_{11}(s)$' on one side of an equation with '$C_{12}H_{22}O_{11}$, aq.' on the other *in a calculation for ΔH*. This would not be justified in a calculation for ΔG, since the reaction (11.33) may involve quite a considerable decrease in free energy; that is why it occurs so spontaneously. In fact, the actual concentration will usually have

---

[1] 'Sucrose, aq.' means sucrose in aqueous solution.

to be taken into account; merely saying 'sucrose, aq.' is inadequate. Where the standard free energy of a reaction is small, reference to equation (11.30) will show that the equilibrium constant is near to unity, and we are therefore dealing with a chemically reversible reaction (see p. 48). Biological systems possess many reactions of this type, since teleologically in operating not too far from their equilibrium position they are able to negotiate energy transfer without large wastage due to thermodynamic irreversibility (see Chapter 4). However, it is just because the *standard* free energy change is small that the *actual* free energy change may be proportionally widely different from it for no other reason than is occasioned by the concentrations differing from unity; in fact simply looking at a table of the standard free energies of biochemical reactions may give quite the wrong impression of the direction in which the reactions proceed in the cell. This is of course only saying that the second term on the right in equation (11.28) may easily swamp the first if the first happens to be numerically small.

Tables of Free Energies

Another problem that concerns the usefulness of free energy data for chemical reactions is how best to tabulate the experimental results available, so that these can be pressed into service for as many different reactions as possible. It is really a question of 'indexing' them and it is not difficult to see that one of the most convenient ways of doing this is to tabulate the affinities of formation of all important compounds from their elements. From these data it is easy to appreciate that the affinity of any other reaction can readily be derived by a process of addition and subtraction of equations as illustrated above for ΔH. Needless to say the affinities of formation of chemical compounds have usually to be derived indirectly as in the case of the ethanol, though there are cases (such as the formation of hydrogen iodide) where they can be measured directly.

Overall Free Energy Change in Respiration

This discussion on the affinities or free energy changes of chemical reactions will be concluded by calculating the overall affinity of the process of respiration, using glucose as a substrate. The chemical equation is

$$C_6H_{12}O_6 + 6O_2 \rightarrow 6CO_2 + 6H_2O.$$

Since there is a considerable change in the free energy of a gas when it changes its pressure $\left(\text{in fact } \Delta F = \Delta G = RT \ln \frac{p_2}{p_1}\right)$ and since there is also an appreciable change when a solute dissolves or diffuses, it will be necessary to specify firstly, the pressures of all gases; and secondly, the concentrations of all solutes. It was seen that neither of these items was commonly of any significance when dealing with ΔH values. Assume at first that the glucose is in the solid state, the water in the liquid, and the gases are at a pressure of one atmosphere. The temperature is uniform throughout all operations (or free energy changes would be physically meaningless) and it will be assumed to be 25°C.

Reference to published tables [1] gives the following data for free energies of formation at 25°C:

(i) $6C(s) + 6H_2(1 \text{ atm.}) + 3O_2(1 \text{ atm.}) \rightarrow C_6H_{12}O_6(s)$; $\Delta G = -217.56$ kcal

(ii) $C(s) + O_2(1 \text{ atm.}) \rightarrow CO_2(1 \text{ atm.})$; $\Delta G = -94.26$ kcal

(iii) $H_2(1 \text{ atm.}) + \tfrac{1}{2}O_2(1 \text{ atm.}) \rightarrow H_2O(l)$; $\Delta G = -56.69$ kcal.

Multiplying (i) by minus one, (ii) by six and (iii) by six and adding the results gives:

(iv) $C_6H_{12}O_6(s) + 6O_2(1 \text{ atm.}) \rightarrow 6CO_2(1 \text{ atm.}) + 6H_2O(l)$;

$$\Delta G = -688.14 \text{ kcal.}$$

This is the answer that was required, but the calculation will be carried a stage further to bring it more into line with normal physiological conditions. This will require firstly, that the glucose is arranged to be in aqueous solution (it will be assumed to be 0·1 molal), and secondly, that the partial pressure of oxygen is altered to 0·2 atmosphere, and that of carbon dioxide to 0·0003 atmosphere.

The gases will be dealt with first. The change in chemical potential[2], or free energy per mole, of a pure gas when the pressure alters from p_1 to p_2 is (Chapter 5, compare equation (5.6))

$$\mu_2 - \mu_1 = RT \ln \frac{p_2}{p_1} \tag{11.34}$$

and this gives the free energy change[3] of the reaction that is required. Thus for the oxygen:

(v) $O_2(0.2 \text{ atm.}) \rightarrow O_2(1 \text{ atm.})$; $\Delta G = RT \ln 5 = 0.95$ kcal

(vi) $CO_2(1 \text{ atm.}) \rightarrow CO_2(0.0003 \text{ atm.})$; $\Delta G = RT \ln (0.0003) = -4.80$ kcal

where $R = 1.987 \times 10^{-3}$ kcal per °C per mole, and $T = 298°$K. These equations have been written the correct way round for adding directly to (iv) after multiplying each by six. It is now necessary to deal with the free energy change when solid glucose dissolves to give a 0·1 molal solution. The

[1] See Krebs, H. A. and Kornberg, H. L. (1957). 'Energy Transformations in Living Matter' (appendix by K. Burton), Springer-Verlag, Berlin.
[2] This is the same in the case of pure substances as the 'free energy per mole'.
[3] See the point made later on the dimensions of the 'free energy change' of a reaction.

important step here is to notice that a saturated solution of glucose at 25°C is approximately 10 molal; hence there is no change in the chemical potential of glucose between the solid and a 10 molal solution:

(vii) $\quad C_6H_{12}O_6(s) \to C_6H_{12}O_6(\text{saturated, aq.}); \qquad \Delta G = \text{zero},$

since by definition a solid substance is in equilibrium with its saturated solution. The problem has therefore been reduced to that of finding the change in free energy when glucose is transferred from a solution of molality 10 to one of molality 0·1, and that can be tackled with equation (11.4) or (11.6). It is in fact given very approximately [1] by:

$$\Delta G = \mu_{\text{glucose}}(0\cdot 1 \text{ molal}) - \mu_{\text{glucose}}(10 \text{ molal})$$
$$= RT \ln\left(\frac{0\cdot 1}{10}\right)$$
$$= -2\cdot 7 \text{ kcal per mole.}$$

Thus:

(viii) $\quad C_6H_{12}O_6(s) \to C_6H_{12}O_6(\text{aq., } 0\cdot 1 \text{ molal}); \qquad \Delta G = -2\cdot 7 \text{ kcal.}$

In passing it might be remarked that equations (vii) and (viii) do not really balance in the usual sense, since there is water on one side and not on the other. However since the water acts only as a medium this does not really matter. What is more important is that it should be realized that expressions like 'the change of free energy of glucose on dilution' or 'the free energy of glucose in solution' which are sometimes met with, and which the statement (viii) might seem to countenance, are apt to be misleading. The free energy of a *system* can be referred to, but strictly not of one chemical species in it; the correct term in the latter case is the chemical potential[2]. It is therefore the chemical potential of the glucose that changes in the situation that has been discussed and this change is related directly to the ΔG of the reaction. One reason why equation (viii) does not justify the use of the expressions noticed above is that in this and similar 'free energy' equations ΔG has the physical dimensions not of free energy, but of *free energy per mole*, just as the chemical potential has. That is one reason why the alternative name, the affinity, is a better one. (It will be remembered that the ΔG of a reaction is equal to $-A$ or $-\dfrac{\partial G}{\partial \xi}$ where ξ is measured in 'composite' moles.) As a consequence of all this, the affinity (or free energy change) of any sort of reaction is always found by adding or subtracting numerical multiples of different chemical potentials (compare equations (11.23) and (11.34)).

[1] On account of the high concentration involved. The '10 molal' is in any case very nominal.
[2] Or alternatively, the partial molar free energy.

The final summation is now made by multiplying equation (v) by six, equation (vi) by six, and equation (viii) by minus one and adding to equation (iv). After cancelling identical terms appearing on both sides the result is

$$C_6H_{12}O_6(\text{aq.}, 0\cdot 1 \text{ molal}) + 6O_2(0\cdot 2 \text{ atm.}) \rightarrow 6CO_2(0\cdot 0003 \text{ atm.}) + 6H_2O(l);$$

$$\Delta G = -708\cdot 24 \text{ kcal.}$$

This figure would naturally have been unchanged had the water emerged as *saturated* vapour[1]; had it been at a partial pressure lower than this the adjustment would have been made on the lines of those for oxygen and carbon dioxide. Incidentally, the chemical potential of an ideal gas is not affected by the presence of other gases; consequently in the calculations for oxygen, carbon dioxide and water vapour the fact that nitrogen or air is also present may be neglected. It is the partial pressure of the gas concerned that matters.

A useful rule to remember, and one which the reader can easily verify from the above calculation, is that if the concentration or pressure of any of the reactants rises, or of the products falls, the value of ΔG moves in the negative direction. That is, a spontaneous reaction is made 'more spontaneous', and vice versa.

Difference Between ΔH and ΔG

Since $G = H - TS$, or under isothermal conditions $\Delta G = \Delta H - T\Delta S$, the difference between the heat of reaction ΔH and the free energy change ΔG lies in the term $T\Delta S$. It so happens that the entropy change in the combustion of glucose is very small; consequently ΔG is very nearly equal to ΔH. In fact ΔH under the conditions of glucose solid, gases at one atmosphere, and water liquid is 673·0 kcal per mole. Needless to say this approximate equality of ΔG and ΔH is not generally true. They can in fact be widely different.

Influence of Temperature

Plant physiologists will probably be content to evaluate affinities at the standard temperature of 25°C at which the data are usually tabulated. Animal physiologists, however, will often require to evaluate these values at higher temperatures. It is not proposed to go into the question in detail of how to refer free energy data to a higher temperature, but the following remarks may be of interest.

It was seen earlier (equation (11.21)) that the rate of change of the chemical

---

[1] It is hardly necessary to point out that this would be quite untrue if ΔH were being calculated instead since the latent heat would be involved.

potential of a component with temperature is equal to the negative partial molar entropy of that component:

$$\frac{\partial \mu_i}{\partial T} = -\frac{\partial S}{\partial n_i} = -\bar{S}_i. \tag{11.35}$$

Now just as the μ's of reactions and products were combined to give the free energy change of a reaction, so the partial molar entropies[1] can be combined in exactly the same way to give the entropy change. Bearing in mind equation (11.35) and the fact that the ΔG value is simply a linear combination of the μ's, it will be appreciated that the relation holds:

$$\frac{\partial (\Delta G)}{\partial T} = -\Delta S \tag{11.36}$$

where ΔS is the entropy change of the reaction just as ΔG is its free energy change, both being dimensionally per mole. The significance of this result is that the sensitivity of the free energy change to temperature depends on whether or not a large entropy change[2] is involved. If the change is small (as in the case of the oxidation of glucose) the free energy change will vary very little with the temperature; if it is large (as in the hydrolysis of starch) it will vary appreciably. If the standard entropy change for the reaction is available it can obviously be used to make an estimate of the affinity at the higher temperature; in fact we shall have

$$\Delta G^{\ominus}(T_2) = \Delta G^{\ominus}(T_1) - \Delta S^{\ominus}(T_1) \times (T_2 - T_1) \tag{11.37}$$

where $\Delta G^{\ominus}(T_2)$ means the standard free energy change at T_2, and so on; data are assumed to be available at T_1. If the standard entropy changes are not accessible they can be calculated if both $\Delta G^{\ominus}$ and $\Delta H^{\ominus}$ values are to hand, since as was just seen a moment ago

$$\Delta G^{\ominus} = \Delta H^{\ominus} - T\Delta S^{\ominus} \text{ or } \Delta S^{\ominus} = \frac{\Delta H^{\ominus} - \Delta G^{\ominus}}{T}. \tag{11.38}$$

Free Energy Change in Transport Processes

This chapter will be concluded with a brief discussion of transport processes to show their analogy with chemical processes, and then a brief summary of the rules for manipulating thermodynamic equations.

If a variable amount of a solute like urea is shaken up vigorously with a mixture of olive oil and water, and then after settling samples of the two liquid phases are withdrawn and analysed, it is found that the ratio of the

---

[1] It will be recalled that μ is the partial molar free energy, so the cases are quite analogous.
[2] A useful point to remember is that where there is an entropy *increase* the free energy yield $(-\Delta G)$ *rises* with temperature and vice versa. Thus if the e.m.f. of a battery rises with temperature the cell reaction has a positive ΔS.

concentrations of the urea in the oil and water is virtually a constant (κ) at any given temperature:

$$\frac{\text{concentration in oil}}{\text{concentration in water}} = \kappa.$$

Of course, it is assumed that equilibrium is achieved and that there is not always sufficient urea present to leave an excess undissolved[1]. The constant κ is called the oil-water partition coefficient, and needless to say its value will depend markedly on how the concentrations are measured, whether in moles per unit volume, molalities, or mole fractions; it will also vary with temperature. In all these things κ follows a close analogy with the equilibrium constant of a chemical reaction.

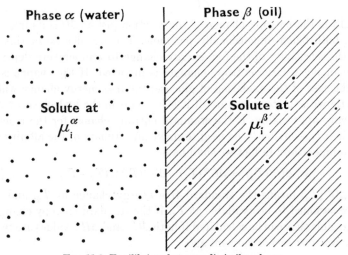

FIG. 11.3. Equilibrium between dissimilar phases.

Now imagine a system containing two immiscible phases α and β (Fig. 11.3), with a solute (i) which can pass between them. Let the chemical potentials of the solute in the two phases be μ_i^α and μ_i^β, the system not being in equilibrium. Then, if a quantity $d\xi$ moles of solute passes from α to β the free energy decrease of the system ($-dG$) can be written

$$-dG = \mu_i^\alpha d\xi - \mu_i^\beta d\xi,$$

or (11.39)

$$-\left(\frac{\partial G}{\partial \xi}\right)_{T,P} = \mu_i^\alpha - \mu_i^\beta,$$

the temperature and pressure remaining unchanged. This equation should be compared with equation (10.22). As in the chemical situation (see p. 143)

[1] Otherwise of course the concentrations are always those at saturation in the two solvents.

we can speak of an affinity (A) for the transport process, or alternatively a free energy change ($-\Delta G$), equal to $-\left(\dfrac{\partial G}{\partial \xi}\right)_{T,P}$; in this case the quantity called the 'extent of reaction' (ξ) is plainly and simply the number of moles of solute which has crossed the boundary. Using the 'free energy change' notation gives

$$\Delta G = -\mu_i^\alpha + \mu_i^\beta. \tag{11.40}$$

Now write $\mu_i = \mu_i^\ominus + RT \ln m_i$, remembering that, since there are two solvents, two senses must be distinguished for m_i and two for $\mu_i^\ominus$. Suppose these are distinguished with superscripts α and β. Then

$$\Delta G = -(\mu_i^{\alpha\ominus} + RT \ln m_i^\alpha) + (\mu_i^{\beta\ominus} + RT \ln m_i^\beta)$$

$$= -\mu_i^{\alpha\ominus} + \mu_i^{\beta\ominus} + RT \ln \frac{m_i^\beta}{m_i^\alpha}. \tag{11.41}$$

The two standard potentials both refer to the same temperature and to unit molality; but since the solvents are different the reader will understand that they cannot simply be equated. Olive oil with a given content of urea per kilogram of oil would not be in equilibrium with an aqueous solution containing the same amount per kilogram of water, since for one thing urea is far more soluble in water.

The unknown standard values are eliminated by the same device as when dealing with chemical reactions. Suppose m_i^β/m_i^α happens to have the value κ. Then the system will be in equilibrium, ΔG will be zero, and thus (inserting these values in equation (11.41)) we shall have

$$0 = -\mu_i^{\alpha\ominus} + \mu_i^{\beta\ominus} + RT \ln \kappa. \tag{11.42}$$

Subtracting equation (11.41) from (11.42) gives

$$-\Delta G = RT \ln \kappa - RT \ln \frac{m_i^\beta}{m_i^\alpha} \tag{11.43}$$

in exact analogy with the chemical formula (11.28). If m_i^α and m_i^β are each made equal to unity then the standard affinity or free energy change is given as

$$A^\ominus = -\Delta G^\ominus = RT \ln \kappa. \tag{11.44}$$

Incidentally, if both phases are aqueous then $\kappa = 1$ and equation (11.43) with (11.40) reduces to the earlier one (11.4).

Summary of Rules for Manipulating Free Energy Equations

(1) In general, the ordinary rules of algebra apply in adding or subtracting equations, the ΔG or ΔH value being regarded as an extra term on one side.

(2) If an equation is multiplied by a number, the ΔG or ΔH value is also multiplied by the same number. Thus in selecting the value from the tables the exact stoichiometrical coefficients to which it relates must be noted carefully (see p. 161).

(3) All equations must ordinarily be 'balanced'.

(4) Reversal of an equation reverses the sign of ΔG or ΔH.

(5) When the equations are finally combined, terms appearing one on each side may be cancelled provided they relate to equal amounts of the same substance; and provided that, in the case of ΔG, the substance has equal chemical potentials, and in the case of ΔH, that it has equal energy contents (or strictly, partial molar enthalpies).

These last two provisos are very important and may be illustrated by the following examples.

(i) $2H_2O(l)$ and $2H_2O$ (saturated vapour): these terms are equal in the amount of substance and may be cancelled in a ΔG calculation, since saturated vapour is in equilibrium with the liquid and therefore has the same chemical potential. (Alternatively a new equation

$$2H_2O(l) \rightarrow 2H_2O(\text{saturated vapour}); \quad \Delta G = 0$$

could be introduced, but this is clearly unnecessary.)

In a ΔH calculation however, these two terms may certainly not be cancelled, since to convert one into the other the latent heat would have to be supplied. It would therefore be necessary to introduce the further equation:

$$2H_2O(l) \rightarrow 2H_2O(\text{saturated vapour}); \quad \Delta H = 20\cdot 94 \text{ kcal.}$$

(ii) CO_2 (1 atm.) and CO_2 (0·0003 atm.): these terms may be cancelled with each other in a ΔH calculation (since both U and H depend only on the temperature for a perfect gas, and here the temperature is constant). However, as was seen previously there is an appreciable free energy change involved consequent on the change in chemical potential. It was necessary therefore to introduce a further equation to allow for this in calculating ΔG.

(iii) D-Glucose (0·1 molal) and L-glucose (0·1 molal): in the case of two optical isomers, since the equilibrium mixture is a 50 : 50 one, the two chemical potentials at equal molalities are equal; and clearly no change in energy content is involved. Thus in both ΔG and ΔH calculations the two terms may be cancelled with one another.

It is hardly necessary to add that all equations must relate to the same temperature, and that this temperature ought to be stated in the calculation.

CHAPTER 12

Dilute Solutions of Electrolytes

> Before he knew where he was, Piglet was in the bath, and Kanga was scrubbing him firmly with a large lathery flannel.
> 'Ow!' cried Piglet.
> 'Don't open the mouth, dear, or the soap goes in,' said Kanga. 'There! What did I tell you?'
>
> WINNIE-THE-POOH
> *A. A. Milne*

What has been taken as the fundamental definition of chemical potential turns upon the maximum amount of work which can be extracted when the species to which the potential belongs moves from one potential to another, uniformity of temperature being presumed. In symbols, if dn_i moles of a species, i moves from a place where its potential is μ_i^α to one where it is μ_i^β, the maximum work the movement can do is given by:

$$\text{maximum work} = (\mu_i^\alpha - \mu_i^\beta)dn_i.$$

This is the maximum *total* work if the movement takes place at constant temperature and volume, and the maximum *useful* work if it takes place at constant temperature and pressure.

This definition of chemical potential was then used to derive a formula showing how it depended on the concentration, and in doing so the concept of osmotic work was used. The tacit assumption was made that the work done by (or against) the osmotic pressure was the only sort of work involved in the movement of substance from one potential to the other. This is no doubt often true, but there are many common cases where work of other sorts is done as well; surface tension is a case in point. But by far the most important example arises in the case of charged particles or ions, for these are subject to electrical forces which naturally do work on them when they undergo movement. It is with these that this chapter will be concerned.

We begin with the experimental fact that when salts[1] are dissolved in water to give a dilute solution they behave as if they split up spontaneously into oppositely charged particles (called ions) which lead independent lives; at least they are independent to a very important extent. What this means is that in discussing electrolyte solutions we often have to think in terms of ions rather than complete salts. Next, it must be recognized that these

---

[1] This refers particularly to what are often called 'strong electrolytes' like KCl or Na_2SO_4.

ions carry charges which are relatively enormous. If a single salt like NaCl is established as a non-uniform solution, its ions will proceed to diffuse in such a way that a uniform concentration ultimately results. During this diffusion the oppositely charged ions behave to some extent independently as we have just remarked, and owing to the fact that the ion of one sign has not quite the same size and mass as the ion of the opposite sign their diffusion rates are different. The lighter, faster ions move ahead of the heavier, slower ones, and this separation of charges sets up a *diffusion potential*. The effect of this potential is clearly to retard the faster ions and speed up the slower ones, with the result that the diffusion potential very quickly ceases to rise, and the salt can then be considered to diffuse as an entity. What is important to notice, however, is that while the diffusion potential may be large enough to be easily measured by suitable physical instruments, the corresponding separation of the ions is much too small to be detected by any known chemical procedure. A potential of one volt may easily arise from an unbalanced ionic concentration of 10^{-15} molality, or less.

Electrical Potential Between Phases

We can therefore speak of an electrical potential difference between two phases of virtually *identical* composition, and this is an important fact for the following reason. Suppose two points (α and β) exist whose electrical potentials are ψ^α and ψ^β. Then what we mean (or wish to mean) by saying that the potential difference between them is ($\psi^\alpha - \psi^\beta$) is that if unit quantity of electricity moves from α to β the amount of work done in the movement is ($\psi^\alpha - \psi^\beta$); or if a quantity, dϵ moves then it is ($\psi^\alpha - \psi^\beta$)dϵ. However, the difficulty is that electricity does not move only in response to electrical forces, and this is proved by such facts as follows. When two different metals (say copper and iron) are touched together they actually *develop* a potential difference[1]. In this case therefore electricity (or rather electrons) undergoes an initial movement where no potential difference initially exists, and it actually proceeds to establish one. Obviously some factor other than electrical force is operative, and *this factor must also contribute work*. It is in fact linked with the entropy, and the system of two metals in contact is formally rather like the Donnan system (p. 94) with two phases in contact, electrons taking the place of ions. Correspondingly more of the electrons in the copper than in the iron are free to move; alternatively we may think of more of those in the iron as structurally bound. Entropy considerations therefore take a hand just as in the Donnan system, and an electrical potential results.

[1] It is not implied that this is a physically meaningful quantity, i.e. actually measurable (cf. the next paragraph on the necessity for chemical identity of the phases, here represented by metals). The point is expressed rather loosely.

It follows that it is only in *chemically identical* phases that this 'other' factor balances out and the movement of charged particles (electrons or ions) is determined *solely by electrical forces*. Consequently it is only in identical phases that the expression $(\psi^\alpha - \psi^\beta)\,d\epsilon$ is related to the work done, so that the quantity of work can be used to define what is meant by $(\psi^\alpha - \psi^\beta)$. Where the phases differ widely in chemical composition (like oil and water solutions) the notion of an electrical potential difference between them is indeed physically quite meaningless; where the phases are not too unlike (one can think of two dilute aqueous solutions), statistical theory can be called upon to help and a meaning given to $(\psi^\alpha - \psi^\beta)$ more or less precise according to circumstances. But in *all* cases the corresponding thermodynamic expression $(\mu_i^\alpha - \mu_i^\beta)\,dn_i$ (where i can stand for electrons as well as for ions or molecules) gives a perfectly sound and unequivocal conceptual foundation. It does this because it takes into consideration the entropy factor as well as the energy one.

Electrochemical Potential of Ions

Suppose we have two aqueous phases sufficiently alike in composition for them to be attributed an electrical potential difference, $(\psi^\alpha - \psi^\beta)$. To digress for a moment, this proviso will mean, in practical terms, that any well-conceived method for measuring $(\psi^\alpha - \psi^\beta)$ will yield virtually the same result: if the phases are too unlike it will be found quite impossible to achieve a satisfactory agreement. The question of how alike the phases must be for us to be able to speak of the electrical potential difference between them can thus be settled on practical grounds, as could the analogous question (p. 3) of whether a system is close enough to equilibrium for a definite temperature or pressure to be attributed to it. Granting the proviso the following observations can be made.

Firstly, if the molality of an ion is m_i this will contribute an amount $RT \ln m_i$ to the chemical potential, just as if it were an ordinary solute. It does this of course by virtue of its osmotic activity. However, it is better to write this contribution $RT \ln \gamma_i m_i$, where γ_i is a correction factor called the 'activity coefficient' (see below).

Secondly, with reference to our fundamental definition of μ, an additional amount of work can be extracted in the case of ions by a suitable mechanism which can harness the electrical forces. If necessary an auxiliary phase could be introduced whose electrical potential was the same as that of α but whose molality was the same as that of β; transfer of the dn_i moles could then be managed in two stages and the osmotic and electrical contributions to the work collected separately[1]. Suppose that the valency of the ion is z (that is,

---

[1] This separability only applies to phases so alike in composition that $(\psi^\alpha - \psi^\beta)$ is well defined; see remarks later.

it carries a charge of z units). Then the amount of electricity associated with $\mathrm{d}n_i$ moles of the ion is $z\mathscr{F}\mathrm{d}n_i$ where $\mathscr{F}$ is the Faraday (96,493 coulombs per gram equivalent); and the electrical work involved is $z\mathscr{F}\mathrm{d}n_i(\psi^\alpha - \psi^\beta)$.

Thus it follows that

$$(\mu_i^\alpha - \mu_i^\beta)\mathrm{d}n_i = RT \ln \frac{\gamma_i^\alpha m_i^\alpha}{\gamma_i^\beta m_i^\beta} \mathrm{d}n_i + z\mathscr{F}(\psi^\alpha - \psi^\beta)\mathrm{d}n_i$$

or

$$\mu_i^\alpha - \mu_i^\beta = RT \ln \frac{\gamma_i^\alpha m_i^\alpha}{\gamma_i^\beta m_i^\beta} + z\mathscr{F}(\psi^\alpha - \psi^\beta). \tag{12.1}$$

Following the same procedure as in the previous chapter one of the phases may be considered to be a standard one with its γm equal to unity and its electrical potential at an arbitrary zero. This means that we may write

$$\mu_i = \mu_i^\ominus + RT \ln \gamma_i m_i + z\mathscr{F}\psi, \tag{12.2}$$

where the chemical potential $\mu_i^\ominus$ refers to the standard phase described.

In order to emphasize the fact that an electrical factor is involved the chemical potentials of ions are often referred to as electrochemical potentials, but the name must not be taken as implying that there is another property of the ion called its chemical potential. In other words it must not be taken to mean that the electrical and 'other' components of the electrochemical potential can always be neatly distinguished, as has just been done. As was seen earlier in the admittedly clear-cut case of an ion in two quite different solvents a meaning cannot be given to $(\psi^\alpha - \psi^\beta)$; so of course the 'electrical' component as a distinguishable part of the whole ceases to have any significance. Even in rather less clear-cut cases it still remains true that $(\psi^\alpha - \psi^\beta)$ is a quantity lacking in the real precision typical of thermodynamics. Therefore it may be taken that the special epithet 'electrochemical' is merely a reminder that a chemical potential is being considered to which electrical forces make an important contribution; and further that the formula (12.2) derived for it, is only valid when considering phases sufficiently alike in composition for $(\psi^\alpha - \psi^\beta)$ to be given a precise enough meaning. Invariably this will mean that the *solvent* species in the two phases must be the same.

Activity and Activity Coefficients

Mention has already been made in a previous chapter (Chapter 10) of the property known as the activity. *Absolute* activities (λ) are related exponentially to the chemical potential by the following equation:

$$\mu_i = RT \ln \lambda_i. \tag{12.3}$$

What is of greater concern however is a property called the *relative* activity (a), and this is related to the difference between the chemical potential and its value in some standard state.[1] In terms of compositions expressed in molalities the relation becomes

$$\mu_i - \mu_i^{\ominus} = RT \ln a_i, \tag{12.4}$$

and a_i can be looked upon as a kind of 'corrected' concentration (cf. equation (11.6)). When dealing with electrolyte solutions it is found much better to use activities instead of uncorrected molalities almost from the start, except for extremely dilute solutions or very approximate calculations. There are two reasons for this. Firstly, the electrostatic forces between ions exert their influences over far greater distances than do the short range interactions of uncharged molecules, and this means that electrolyte solutions depart much sooner from ideality. Secondly, the law of electrostatic force is known exactly, so that it becomes possible to calculate with fair precision the factor by which the molality must be multiplied to yield the activity. This factor is called the activity coefficient, and is given by

$$\gamma_i = \frac{a_i}{m_i} \quad \text{or} \quad a_i = \gamma_i m_i. \tag{12.5}$$

In the following discussion sometimes activities will be used, sometimes molalities and activity coefficients, and sometimes for simplicity plain molalities. The reader should remember that the use of a_i or $\gamma_i m_i$ yields exact relations, and that the use of m_i alone, especially with electrolytes, is an approximation. Some of the properties of activity coefficients for electrolytes will be discussed later on.

CHEMICAL POTENTIAL OF A SALT

Situations frequently arise in which, in the very nature of the case, salts have to move as entities. For instance, where a single salt is present in a non-uniform solution this is obviously true, since any appreciable separation of ions as diffusion proceeds would set up electrical potentials too great to be sustained without violent discharges. In situations like this we speak of the chemical potential of the *salt*. How is this related to the electrochemical potential of the constituent ions? If the reader reflects that osmotic potential is a colligative property (and therefore additive) it will be obvious that the osmotic component of work involved in a transfer of salt is the sum of the components which could be attributed to the ions individually, and a similar statement is true of the electrical components. Thus the chemical potential of a salt is simply the sum of the electrochemical potentials of its component ions. For a salt with two univalent ions:

$$\mu_{MA} = \mu_{M^+} + \mu_{A^-} \tag{12.6}$$

[1] It is tacitly assumed here that the standard state is at the same electrical potential as the other.

where MA is the salt, and M⁺, A⁻ are the ions into which it dissociates (subsequently the subscripts M and A will be used without the signs where confusion is unlikely). If this equation is combined with (12.2) we have

$$\mu_{MA} = \mu_M^\ominus + RT \ln \gamma_M m_M + \mathscr{F}\psi + \mu_A^\ominus + RT \ln \gamma_A m_A - \mathscr{F}\psi$$
$$= (\mu_M^\ominus + \mu_A^\ominus) + RT \ln (\gamma_M \gamma_A\, m_M m_A). \tag{12.7}$$

This equation illustrates several important points. In the first place, the electrical terms have disappeared. This is a consequence of the fact that the salt is electrically uncharged, and it obviously holds for salts with ions of valencies other than the simple 1 : 1 values taken for the sake of illustration. For this reason a salt is attributed a plain *chemical* potential. In the second place, the activity coefficients γ_M, γ_A occur as a product $\gamma_M \gamma_A$. Such a product[1] of activity coefficients always turns up in equations which are 'physically meaningful' (see p. 156), except that where a salt molecule dissociates into more than one of a particular ion, the activity coefficient of that ion is raised to the corresponding power. Thus in the case of K₂S the product $\gamma_{K^+}^2 \gamma_{S^{2-}}$ will occur, with FeCl₃ the product $\gamma_{Fe^{3+}} \gamma_{Cl^-}^3$. In practice, activity coefficients are obtained by measuring such things as freezing point depressions, vapour pressure lowerings or the e.m.f. of cells, and comparing them with ideal values. Naturally experiments have to be carried out with whole salts since a solution of a single ion would be unthinkable; it is a consequence of this that all that can ever be measured are the aforementioned *products* of activity coefficients, not the values for single species of ions. This leads naturally to the introduction of a 'mean activity coefficient' ($\bar{\gamma}_{MA}$) for the salt, illustrated by such examples as

$$\bar{\gamma}_{MA}^2 = \gamma_{M^+} \gamma_{A^-} \quad \text{(compare KCl)}$$

or

$$\bar{\gamma}_{MA}^4 = \gamma_{M^{3+}} \gamma_{A^-}^3 \quad \text{(compare FeCl}_3\text{)}$$

where the index on the left is the sum of those on the right, thus keeping the equations dimensionally consistent.

In the third place, if $\gamma_M m_M, \gamma_A m_A$ are each put equal to unity in equation (12.7) and correspondingly μ_{MA} is written equal to $\mu_{MA}^\ominus$ as was done analogously with μ_i, then the result follows

$$\mu_{MA}^\ominus = \mu_M^\ominus + \mu_A^\ominus. \tag{12.8}$$

Finally for a salt MA[2]

$$\mu_{MA} = \mu_{MA}^\ominus + RT \ln (\bar{\gamma}_{MA}^2 m_M m_A). \tag{12.9}$$

---

[1] It will be a quotient where the associated ions have the same sign, as in some electrochemical cells.

[2] The equation is written for a 1:1 salt. For salts with ions of other valencies it is quite easy to modify it.

It should be noted that in deriving this equation the ions were imagined to be moving from phase α to phase β in equal numbers, thus preserving electrical balance; but it was not assumed that in either phase the molalities m_M and m_A were equal. In a mixture of salts they obviously need not be.

Some Properties of the Activity Coefficients

It has already been seen that activity coefficients for single ionic species are not measurable, and that therefore mean activity coefficients are used. Suppose we have a solution containing a large number of different anions and cations. None of the anions can be said to 'belong' to one cation more than to another; put in another way, any salt may be regarded as being present whose constituent ions are there. Any such possible salt will have its own mean activity coefficient, and this will depend not only on the concentrations of its constituent ions but also on those of other quite unrelated ions, since all exert quite general electrical forces. In this connection a quantity appears which is called the *ionic strength* (I) of the solution. This is a measure of its total content of electrically charged particles, and is defined as

$$I = \tfrac{1}{2} \Sigma z_i^2 m_i. \qquad (12.10)$$

Since the valency z appears as a square, anions and cations alike make *positive* contributions to I. Further, and for the same reason, di- or trivalent ions contribute much more than their aggregate charge would suggest. For a solitary 1:1 salt like KCl the value of I is clearly simply m, the molality of the salt; for a solitary 2:2 salt like $MgSO_4$ it is $4m$. Statistical theory predicts that for a dilute solution the mean activity coefficient of a salt is given approximately by

$$\ln \bar{\gamma} = -c z_1 z_2 \sqrt{I}$$

or

$$\log_{10} \bar{\gamma} = -d z_1 z_2 \sqrt{I}, \qquad (12.11)$$

where d is approximately 0·51 molal$^{-\frac{1}{2}}$ for water at 25°C[1], and z_1, z_2 are the valencies of its two ions, both here regarded as positive numbers. As an example of the use of this formula we can calculate the activity coefficient for a 1:1 salt like NaCl present alone at a strength of 0·05 molal. In this case

$$I = \tfrac{1}{2}[1^2 \times 0.05 + 1^2 \times 0.05] = 0.05 \text{ molal},$$
$$\sqrt{I} = \sqrt{0.05} = 0.224 \text{ molal}^{\frac{1}{2}}.$$
$$\log_{10} \gamma = -0.51 \times 1 \times 1 \times 0.224$$
$$= -0.114$$

giving $\gamma = 0.77$.

---

[1] The corresponding value of c is 1·17.

This compares with a value obtained from e.m.f. measurements of about 0·82. For a solution of molality 0·01 the calculated value is 0·890 compared with a measured one of 0·903; and needless to say as the ionic strength becomes less γ tends to move nearer and nearer to unity. For extremely dilute solutions it can be taken as unity, but the criterion of diluteness is the *total* ionic strength and not the molality of the particular salt under consideration. For instance, the mean activity coefficient of NaCl is 0·903 in a solution of molality 0·01 when no other salts are present; but if 0·04 molal KBr is also present it falls to 0·82. 'Foreign' di- and trivalent ions will naturally be very potent in this connection, owing to their large contribution to I.

The Donnan Equilibrium

We are now in a position to discuss the Donnan equilibrium. This is an equilibrium which occurs very frequently indeed in biological systems. It arises when there are two aqueous phases in contact, in one of which there is an ion which is held confined to that phase. It may be confined because there

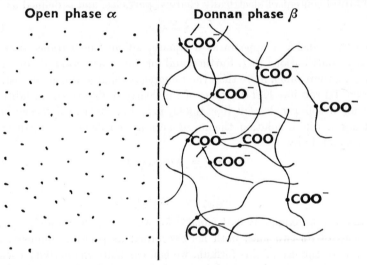

Fig. 12.1. A simple Donnan system.

is a membrane separating the two phases which it cannot penetrate; alternatively it may be bound to the macromolecular framework of one phase quite apart from any membrane. Many cell vacuoles are rich in colloid material, and such vacuoles would serve as an illustration of the first type of Donnan phase. As an example of the second type we can think of the equilibrium between ions in the soil solution and those in the colloids of the soil humus or the pectins of the cell wall.

Imagine such a two-phase system (Fig. 12.1) with an open phase α, and a Donnan phase β, consisting of a macromolecular framework carrying numerous negatively charged groups, such as $-COO^-$. The framework might have had positively charged groups; again, there might have been a membrane separating the phases, but in both cases the following theory would hold in all essentials. The phases are both aqueous and there are present a wide assortment of mobile ions of both signs, for instance Na^+, K^+, Ca^{2+}, Cl^-, NO_3^- and SO_4^{2-}, to make matters definite. The problem is to determine the equilibrium distribution of all the ions between the two phases, and we start with the assumption suggested by many observations (as earlier noticed) that the ions can move individually across the boundary; that is, as ions and not exclusively as integral salts. With this assumption it can be said at once that at thermodynamic equilibrium the electrochemical potentials of *any* mobile ion in the two phases are equal. Thus such equations arise as

$$\mu_{Na}^{\alpha} = \mu_{Na}^{\beta} \quad \text{or} \quad \mu_{Na}^{\alpha} - \mu_{Na}^{\beta} = 0 \tag{12.12}$$

$$\mu_{Ca}^{\alpha} = \mu_{Ca}^{\beta}$$

$$\mu_{SO_4}^{\alpha} = \mu_{SO_4}^{\beta}$$

and so on. In general, it must be assumed that an electrical potential difference $(\psi^{\alpha} - \psi^{\beta})$ develops between the phases. Thus if the general formula (12.1) or (12.2) is introduced into the relation (12.12) then we find

$$\mu_{Na}^{\alpha} - \mu_{Na}^{\beta} = 0 = RT \ln \left(\frac{\gamma_{Na}^{\alpha} m_{Na}^{\alpha}}{\gamma_{Na}^{\beta} m_{Na}^{\beta}} \right) + z\mathscr{F}(\psi^{\alpha} - \psi^{\beta})$$

or since $z = +1$ for the sodium ion,

$$RT \ln \left(\frac{\gamma_{Na}^{\alpha} m_{Na}^{\alpha}}{\gamma_{Na}^{\beta} m_{Na}^{\beta}} \right) = -\mathscr{F}(\psi^{\alpha} - \psi^{\beta}). \tag{12.13}$$

Equations similar to (12.13) will hold for all the other univalent cations. For univalent anions, z has the value -1, and this can be allowed for (leaving unaltered the right hand side) by turning upside down the composite fraction whose logarithm has to be taken. For divalent ions, $z = \pm 2$ and again the right hand side can be kept unaltered by taking the square root of this same composite fraction. In this way, since the right hand side of (12.13) is common to all the results it is possible to deduce

$$\left(\frac{a_{Na}^{\alpha}}{a_{Na}^{\beta}} \right) = \left(\frac{a_{K}^{\alpha}}{a_{K}^{\beta}} \right) = \left(\frac{a_{Ca}^{\alpha}}{a_{Ca}^{\beta}} \right)^{\frac{1}{2}} = \left(\frac{a_{Fe}^{\alpha}}{a_{Fe}^{\beta}} \right)^{\frac{1}{3}} = \dots$$

$$= \left(\frac{a_{Cl}^{\beta}}{a_{Cl}^{\alpha}} \right) = \left(\frac{a_{NO_3}^{\beta}}{a_{NO_3}^{\alpha}} \right) = \left(\frac{a_{SO_4}^{\beta}}{a_{SO_4}^{\alpha}} \right)^{\frac{1}{2}} = \dots \tag{12.14}$$

where for clarity the activity $(a = \gamma m)$ has been used, and the multiplier RT and the logarithmic form have been got rid of.

If the relations (12.14) are scrutinized it is noticed that whereas all the cations have the α in the numerator and the β in the denominator, with the anions it is the other way round. Further an index appears which is the reciprocal of the valency. Suppose a 'possible' salt is considered by selecting the fraction for any one of the cations present (say Na)[1] and for any one of the anions (say Cl). This gives

$$\left(\frac{a^\alpha_{Na}}{a^\beta_{Na}}\right) = \left(\frac{a^\beta_{Cl}}{a^\alpha_{Cl}}\right)$$

or cross multiplying,

$$a^\alpha_{Na} a^\alpha_{Cl} = a^\beta_{Na} a^\beta_{Cl} \qquad (12.15)$$

and this is an example of *the fundamental relationship of the Donnan equilibrium*. Expressed in words it means that the product of the concentrations (strictly, the activities) of *any* mobile cation and *any* mobile anion present has the same value in both phases. This statement is worded to apply to univalent ions; for divalent or multivalent ones it is a little different. Thus for Ca^{2+} and NO_3^- from (12.14)

$$\left(\frac{a^\alpha_{Ca}}{a^\beta_{Ca}}\right)^{\frac{1}{2}} = \left(\frac{a^\beta_{NO_3}}{a^\alpha_{NO_3}}\right)$$

or

$$a^\alpha_{Ca}(a^\alpha_{NO_3})^2 = a^\beta_{Ca}(a^\beta_{NO_3})^2. \qquad (12.16)$$

It can easily be verified that in terms of molalities and mean activity coefficients, equation (12.16) (taken as a typical example) becomes

$$(\bar{\gamma}^\alpha_{Ca(NO_3)_2})^3 m^\alpha_{Ca}(m^\alpha_{NO_3})^2 = (\bar{\gamma}^\beta_{Ca(NO_3)_2})^3 m^\beta_{Ca}(m^\beta_{NO_3})^2 \qquad (12.17)$$

where the index of the mean activity coefficient $\bar{\gamma}_{Ca(NO_3)_2}$ is the total number of ions into which the salt molecule dissociates.

For the biologist an important point to notice from (12.14) is that with very dilute solutions (in which m_i can be written for a_i) the 'accumulation ratio' (m^β_i/m^α_i) is *the same for all cations of the same valency*, and for equivalent anions has the reciprocal value. For ions of twice the valency these ratios must be squared, and so on. As a concrete example, if a Donnan phase has accumulated sodium to ten times the concentration in the (highly dilute) external medium it will also have accumulated potassium in the same ratio, and the pH will be one unit lower inside the phase; but chloride or hydroxyl will be ten times more abundant outside. Calcium on the other hand will be one hundred times more abundant inside, and ferric iron one thousand times; while sulphate will be one hundred times more concentrated in the outer phase. All this, of course, assumes that these ions are present, and that thermodynamic equilibrium obtains.

[1] For clarity the + and − signs will often be omitted.

Origin of the Electrical Potential Difference

So far it has merely been assumed that an equilibrium potential difference ($\psi^\alpha - \psi^\beta$) exists and that ions distribute themselves passively under its influence. The relationships (12.14) arise because this electrical influence is common to all ions, and of course they hold no matter how the potential originates. In a Donnan system this membrane potential, as it is often called, arises because at least one of the ions present is non-mobile. Were it not for this the potential would vanish; and because of it every mobile ion present has its distribution affected. The magnitude of the potential will be discussed later.

The Accumulation Ratio

In order to evaluate the potential in simple cases, and incidentally to find the accumulation ratio, something needs to be known about the concentration of the Donnan charges relative to that of the mobile ions. This concentration can be measured in equivalents, for obviously the valency of the Donnan ions

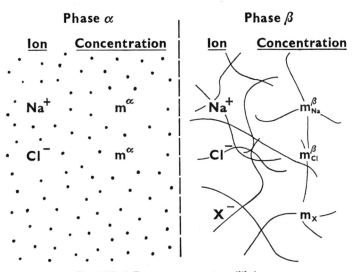

Fig. 12.2. A Donnan system at equilibrium.

is immaterial, if indeed it has any meaning in many cases. Imagine a Donnan system with an open phase α containing sodium chloride of molality m^α. For this simple salt, m^α will also be the molality of each of its ions. Suppose that the Donnan phase β contains fixed negative charges X of equivalent concentration m_X. These charges must be accompanied by balancing cations;

for simplicity suppose these are sodium ions. At equilibrium let the concentrations be as shown (Fig. 12.2). In phase α since electrical neutrality must be maintained both ions naturally have the same molality m^α. In phase β electrical neutrality requires that

$$m^\beta_{Na} = m^\beta_{Cl} + m_X. \tag{12.18}$$

If m^α is regarded as being experimentally fixed (as is often the case) and m_X as hypothetically fixed, there remain two unknowns (the two m^β's), so another equation is needed to enable the problem to be solved. This is the equilibrium condition (12.15) previously found:

$$m^\alpha_{Na} m^\alpha_{Cl} = m^\beta_{Na} m^\beta_{Cl}, \tag{12.19}$$

where for simplicity molalities are used instead of activities. Since $m^\alpha_{Na} = m^\alpha_{Cl} = m^\alpha$, and substituting $m^\beta_{Cl} = m^\beta_{Na} - m_X$ (from (12.18)) into (12.19) we have

$$(m^\alpha)^2 = m^\beta_{Na}(m^\beta_{Na} - m_X). \tag{12.20}$$

This enables m^β_{Na} to be found in terms of m^α and m_X; but before solving this equation it can be cast into a better form. We call $m^\beta_{Na}/m^\alpha_{Na}$ the 'accumulation ratio' for sodium and write it ρ. Further, we call m_X/m_α the 'relative Donnan strength' and write it $\mathscr{D}$. With these symbols the equation becomes

$$\rho(\rho - \mathscr{D}) = 1$$

or

$$\rho^2 - \mathscr{D}\rho - 1 = 0, \tag{12.21}$$

from which

$$\rho_{Na} = \sqrt{\left(\frac{\mathscr{D}}{2}\right)^2 + 1} + \frac{\mathscr{D}}{2} \tag{12.22}$$

while a similar treatment for ρ_{Cl} easily yields

$$\rho_{Cl} = \sqrt{\left(\frac{\mathscr{D}}{2}\right)^2 + 1} - \frac{\mathscr{D}}{2}. \tag{12.23}$$

Clearly $\rho_{Na}\rho_{Cl} = 1$, in agreement with (12.19).

Variation of Accumulation Ratio with Concentration

It is interesting to notice some quantitative consequences of these equations. The accumulation ratio depends only on $\mathscr{D} = m_X/m^\alpha$, the 'relative Donnan strength', and is therefore high when either the external concentration (m^α) is low, or when the concentration of Donnan charges (m_X) is high. When $\mathscr{D}$ is very large the accumulation ratio for the ions of opposite sign is almost equal to it, and for the ions of the same sign is the reciprocal of $\mathscr{D}$.

When $\mathscr{D}$ is unity these accumulation ratios become 1·62 and 0·62 respectively (again their product is one). The variation of the accumulation ratio with relative Donnan strength is shown in Fig. 12.3.

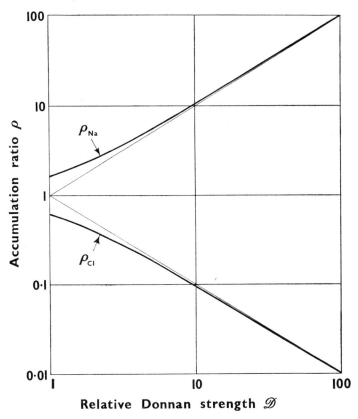

Fig. 12.3. Accumulation ratios for oppositely charged (Na^+) and similarly charged (Cl^-) ions in a Donnan system.

VARIATION OF ACCUMULATION RATIO WITH VALENCY

In the case of a divalent ion (say Ca^{2+}) associated with univalent chloride equation (12.18) becomes

$$2m_{Ca}^\beta = m_{Cl}^\beta + m_X, \tag{12.24}$$

and (12.19) becomes

$$m_{Ca}^\alpha (m_{Cl}^\alpha)^2 = m_{Ca}^\beta (m_{Cl}^\beta)^2, \tag{12.25}$$

with in addition,

$$2m_{Ca}^\alpha = m_{Cl}^\alpha. \tag{12.26}$$

If these equations are manipulated in an analogous way to the previous ones the cubic equation results:

$$\rho(\rho-\mathscr{D})^2 = 1 \quad \text{(cf. (12.21))} \tag{12.27}$$

where ρ is now $m_{Ca}^\beta/m_{Ca}^\alpha$ and $\mathscr{D}$ is $\dfrac{m_X}{2m_{Ca}^\alpha} = \dfrac{m_X}{m_{Cl}^\alpha}$. It is not proposed to discuss this equation further however. The reader interested in plotting the variation of ρ with $\mathscr{D}$ can easily solve it graphically. It is also a simple matter to solve the problem for a trivalent ion. These remarks apply only to the case where single salts are present; where there is a mixture of salts the problem becomes algebraically rather complicated.

Magnitude of the Donnan Electrical Potential

If the accumulation ratio of the ions in a Donnan system is known, equation (12.13) (with (12.14) if necessary) gives the electrical potential $(\psi^\alpha-\psi^\beta)$, or vice versa. However, the situation occasionally arises in which the concentration of the Donnan ions is given, and that of the mobile ions in the external phase, and the resulting membrane potential is required. Where only a single pair of mobile ions is present the matter is quite a simple one. Equation (12.22) gives the accumulation ratio in the case of sodium chloride, and this has only to be combined with equation (12.13) to obtain the required result. Equation (12.13) can in fact be re-written using ρ_{Na} (note the change of sign[1])

$$RT \ln \rho_{Na} = +\mathscr{F}(\psi^\alpha-\psi^\beta)$$

or

$$\psi^\alpha-\psi^\beta = \frac{RT}{\mathscr{F}} \ln \rho_{Na}. \tag{12.28}$$

For an ion of valency z equation (12.28) becomes

$$\psi^\alpha-\psi^\beta = \frac{RT}{z\mathscr{F}} \ln \rho_{ion} \tag{12.29}$$

but this result, like equation (12.28), assumes that the activity coefficients may be taken as equal in the two phases, which will certainly not be true if they differ widely in ionic strength.

A numerical estimate can now be made of the Donnan potential in a typical case. Suppose that the concentration of the (negative) Donnan charges is 0·1 equivalents per kilogram of water (approximately 0·1 normal). Suppose further that the open phase contains only a simple 1:1 salt like

[1] Since the molality fraction has been inverted.

NaCl or KNO_3 at a concentration of 0·001 molal. Then the relative Donnan strength is $0·1/0·001 = 100$ and the accumulation ratio for the cation is

$$\rho_M = \sqrt{\left(\frac{100}{2}\right)^2 + 1} + \frac{100}{2} = 100·01$$

from (12.22). Further, taking $\mathscr{F} = 96{,}493$ coulombs per gram equivalent $R = 8·315$ joules per °C per mole, $T = 25°C = 298°K$, we obtain the result

$$(\psi^\alpha - \psi^\beta) = \frac{8·315 \times 298}{96{,}493} \ln 100·01 \text{ volts}$$

$$= 118 \text{ millivolts}.$$

If the external concentration were ten times higher, or the Donnan concentration one tenth as high (reducing $\mathscr{D}$ to the square root of its previous value), it can easily be confirmed that the potential would work out at very nearly one half of this, i.e. 59 millivolts. A simple rule makes it easy to remember the sign of the potential; *the Donnan phase always has the sign of its fixed charges*. In the present example these have been taken as negative. Consequently the inner phase is always negative with respect to the outer, and $(\psi^\alpha - \psi^\beta)$ has a positive value. The only way to reverse it would be to make the fixed charges positive.

Effect of Departure from Ideality on Accumulation Ratios

In the previous discussion it was assumed that the activity coefficients were the same for any ion in the two phases so that they could be neglected. To what extent is this an approximation? The sort of practical situation which might arise for the physiologist, and in which a knowledge of this point would be valuable, is the following. Measurements have been made of the potential difference $(\psi^\alpha - \psi^\beta) = \Delta\psi$ between an experimental medium and what is supposed to be a Donnan phase of unknown constitution. Further, the accumulation ratios of both sodium (ρ_{Na}) and calcium (ρ_{Ca}) have been found, but their ratio to each other does not agree with that given by suitably combining the simple equations (12.28) and (12.29), viz.

$$\frac{\rho_{Ca}}{\rho_{Na}} = \exp\left(\frac{\mathscr{F}\Delta\psi}{RT}\right). \tag{12.30}$$

To what extent is the variation of the activity coefficients from unity capable of explaining this result, and so allowing the supposition to be retained that the system is a Donnan one?

This problem can be tackled in the following way. Equation (12.30) in its exact form would require ρ_{Ca} (which is equal to $m_{Ca}^\beta/m_{Ca}^\alpha$) to be replaced by $\gamma_{Ca}^\beta/\gamma_{Ca}^\alpha \cdot \rho_{Ca}$, and similarly for ρ_{Na}. If for simplicity the γ^αs are each assumed to be unity (this involves only a minor loss of generality), the approximation (12.30) is equivalent to taking $\gamma^\beta_{Ca}/\gamma^\beta_{Na}$ as unity. Now for a solution which is reasonably dilute (compare (12.11))[1]

$$\ln \gamma_i = -cz^2\sqrt{I} \qquad (12.31)$$

where c is about 1·17 molal$^{-\frac{1}{2}}$ at 25°C, and z is the valency of the ion i. If the Donnan phase has an ionic strength I of 0·1 molal (which would not seem unreasonable) then for the two ions

$$\gamma^\beta_{Na} = \exp(-1\cdot17 \times 1^2 \times \sqrt{0\cdot1})$$
$$\gamma^\beta_{Ca} = \exp(-1\cdot17 \times 2^2 \times \sqrt{0\cdot1})$$

and their ratio,

$$\gamma^\beta_{Na}/\gamma^\beta_{Ca} = \exp(1\cdot17 \times 3 \times \sqrt{0\cdot1})$$
$$= e^{1\cdot11} = 3\cdot0.$$

Put into words, this means that if it was expected on a basis of the simple theory that calcium would be accumulated from a very dilute solution, to a ratio ten times higher than sodium, it is not unduly surprising if its accumulation ratio is found to be $10 \times 3\cdot0$ or thirty times higher. Such a result could easily be explained, without abandoning the supposition of a simple Donnan system, on a basis of lowered activity coefficients in the inner phase. It has to be remembered that higher valency ions suffer a larger reduction in activity coefficient than univalent ones in a medium of given ionic strength (cf. the z^2 in equation (12.31)); consequently they have to make up for this by increasing their concentration. Since the Donnan phase may be ionically appreciably stronger than the external phase, lowered activity coefficients will increase the accumulation ratio and higher valency ions will be accumulated more than would otherwise have been expected.

Further, it is obvious from the foregoing discussion that if the potential difference, $\Delta\psi$ has been measured together with the actual accumulation ratios for a number of different ions, several independent estimates of the ionic strength (I^β) of the Donnan phase can be made, using equation (12.29) in its exact form

$$\Delta\psi = \frac{RT}{z\mathscr{F}}\ln\left(\frac{\gamma^\beta_i}{\gamma^\alpha_i}\rho_i\right) \qquad (12.32)$$

to calculate γ^β_i (γ^α_i can be easily found from the known composition of the outer phase). Then using equation (12.31) I^β can also be calculated. By this means not only is the value of I^β found, but also confirmation (or otherwise) is obtained of the correctness of the assumption that one is dealing with a Donnan system. Of course, the system may behave as a Donnan one for some ions, but not for others if these are subject to active transport.

[1] Equation (12.31) differs from (12.11) in that it concerns the activity coefficient of a single *ion*, not the mean activity coefficient of a *salt*.

Osmotic Unbalance

There is one further aspect of a Donnan system which has not hitherto been mentioned, but which is nevertheless of some importance. When a Donnan phase is put in contact with a salt solution a movement of its ions into the Donnan phase takes place until they have each achieved equality of electrochemical potential in the two phases. This process does not guarantee that the *water* is left with equality of its chemical potential. In general, this will certainly not be the case. Consequently, unless the phases are separated by a rather peculiar membrane permeable to ions but impermeable to water, a movement of water will occur simultaneously with that of the ions. It is very probable that an active, electro-osmotic movement of water may also take place, but this is of a transient nature in a system such as the one under consideration, and something quite different is referred to.

Consider the case dealt with previously where sodium and chloride are the only mobile ions present. The equilibrium state depicted in Fig. 12.2 can be represented in a rather modified form (Fig. 12.4), in which ρ refers to the accumulation ratio of the sodium, given by equation (12.22):

$$\rho = \sqrt{\left(\frac{\mathscr{D}}{2}\right)^2 + 1} + \frac{\mathscr{D}}{2} \tag{12.33}$$

and the corresponding value for chloride is given by equation (12.23):

$$\frac{1}{\rho} = \sqrt{\left(\frac{\mathscr{D}}{2}\right)^2 + 1} - \frac{\mathscr{D}}{2}. \tag{12.34}$$

Since it is osmotic effects that are of interest here and not electrical ones, it is not the concentration of the Donnan *charges* which is relevant but rather that of the macromolecules, and this is usually negligible. The total molalities in the two phases are summed to find the osmotic unbalance between them, and it is apparent that *the Donnan phase always has a higher osmotic potential than the open phase*, since $\left(\rho + \dfrac{1}{\rho}\right)$ is necessarily greater than two[1]. An osmotic pressure must therefore develop within it if equilibrium is to be achieved. However, there is a further question. Is the osmotic differential greater when the Donnan phase is in contact with salt solution than when it is in contact with water? In the latter case the osmotic differential is due almost entirely to the mobile ions partnering the Donnan charges, and the molality of these, if they are univalent, is m_X. In the former case, when salt is present, the osmotic differential is $\left\{\left(\rho + \dfrac{1}{\rho}\right)m^\alpha - 2m^\alpha\right\}$, and it is therefore a question of whether this expression is greater or less than m_X. With

---
[1] The value of ρ cannot be *unity* if a Donnan phase, however weak, is present.

the help of equations (12.33) and (12.34), and bearing in mind that $\mathscr{D} = m_X/m^\alpha$ it can easily be shown that the net osmotic potential, or the osmotic differential, is progressively lowered with increasing concentrations of salt, until it finally becomes zero with strong solutions. This assumes that the molality (m_X) of the Donnan charges remains fixed. However, if the salt is replaced with a strong acid or alkali the ionization (and so the charge density) of the

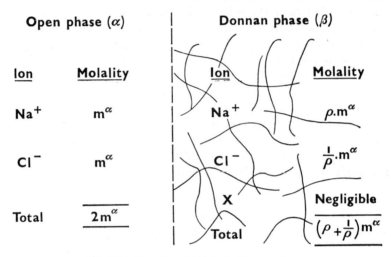

FIG. 12.4. Osmotic potential of a Donnan phase.

Donnan groups may well vary with the concentration of mobile ions, and the behaviour becomes more complex, showing maxima and minima. This is a well-known feature of the influence of pH on the swelling of gelatine and similar proteins. This phenomenon will not, however, be discussed further.

INFLUENCE OF PRESSURE ON ION DISTRIBUTION

It was seen that if a Donnan system is to attain total equilibrium a pressure must develop in the Donnan phase, a pressure which is greater (with fixed concentration of the Donnan charges) the more dilute the solution of mobile ions is. As we saw in the last chapter the pressure on a phase influences the chemical potential of all species present; thus it becomes a matter of interest to enquire whether a pressure difference between phases α and β will alter, to any appreciable extent, the equations for ion distribution already derived. A similar problem has already been investigated for the case of ordinary solutes (Chapter 11), when it was found that the effect depends on the partial molar volume ($\bar{V}_i$) of the solute, and that if $\bar{V}_i = 0.2$ litre an excess pressure in one phase of ten atmospheres will reduce the equilibrium concentration of solute in that phase by about 8%. The present problem is very similar, but as an instructive exercise it will be treated in a slightly different way and the result obtained in a different setting. We shall find out what increment in the membrane potential ($\psi^\alpha - \psi^\beta$) will balance an excess pressure of ten atmospheres on one phase, the concentrations of the ion in question retaining the values already derived assuming zero pressure difference (see equation (12.22) for instance). The reader will notice that in dealing with *ions* the excess pressure can be envisaged as 'balanced' not only by a change in concentration in the two phases (the electrical potential remaining as before), but also by a change in the membrane potential (the concentration remaining as before).

We may start from the relation derived earlier (11.4)

$$\frac{\partial \mu_i}{\partial P} = \overline{V}_i, \tag{12.35}$$

the differentiation being at constant temperature, electrical potential and composition. If the phases are regarded as incompressible (a point implicit in the earlier treatment in Chapter 11), $\overline{V}_i$ will be constant with respect to P and equation (12.35) can be integrated at once to give

$$\mu_i = P\overline{V}_i + \text{(terms independent of } P\text{)}. \tag{12.36}$$

This shows that the effect of pressure can be accommodated by including a new term $(P\overline{V}_i)$ in the expression for μ_i. Thus a more complete statement than before is (cf. equation (12.2))

$$\mu_i = \mu_i^\ominus + RT \ln \gamma_i m_i + z_i \mathscr{F} \psi + P\overline{V}_i, \tag{12.37}$$

where $\mu_i^\ominus$ is now the electrochemical potential at unit activity ($\gamma_i m_i = 1$), zero electrical potential ($\psi = 0$) and zero pressure ($P = 0$). This equation is now applied to two phases α and β in equilibrium, in the usual way. This gives the result

$$RT \ln \gamma_i^\alpha m_i^\alpha + z_i \mathscr{F} \psi^\alpha + P^\alpha \overline{V}_i = RT \ln \gamma_i^\beta m_i^\beta + z_i \mathscr{F} \psi^\beta + P^\beta \overline{V}_i$$

or

$$z_i \mathscr{F} \Delta \psi = -\overline{V}_i \Delta P + RT \ln \left(\frac{\gamma_i^\beta m_i^\beta}{\gamma_i^\alpha m_i^\alpha} \right), \tag{12.38}$$

where $\Delta \psi = \psi^\alpha - \psi^\beta$ and $\Delta P = P^\alpha - P^\beta$. Now suppose that when ΔP is zero the value of $\Delta \psi$ which balances this equation is $\Delta \psi_0$. This is of course the value which has hitherto been dealt with (equation (12.29)). Therefore writing

$$z_i \mathscr{F} \Delta \psi_0 = RT \ln \left(\frac{\gamma_i^\beta m_i^\beta}{\gamma_i^\alpha m_i^\alpha} \right),$$

and subtracting this from equation (12.38), we find

$$z_i \mathscr{F} (\Delta \psi - \Delta \psi_0) = -\overline{V}_i \Delta P. \tag{12.39}$$

It can readily be seen that this result gives the change in membrane potential ($\Delta\psi - \Delta\psi_0$) which must be applied concurrently with the development of a pressure difference (ΔP) to hold the concentrations in equilibrium at their original values. It leads therefore to the result we require[1].

The magnitude of the effect can be seen in a typical case. Consider the sodium ion, and suppose that it moves with a tightly bound shell of ten water molecules. Again $\bar{V}_i$ may be taken as approximately 0·2 litre. Further, $z = 1$, $\mathscr{F} = 96{,}493$ coulombs per mole, and $\Delta P = 10$ atmospheres (say). The product $\bar{V}_i \Delta P$ is $0·2 \times 10$ litre atmospheres, or $0·2 \times 10 \times 101·3$ joules, so that finally we have

$$\Delta\psi - \Delta\psi_0 = -\frac{0·2 \times 10 \times 101·3}{1 \times 96{,}493} \text{ volts}$$

$$= -2·1 \text{ millivolts}.$$

Since the potentials frequently existing across plant and animal membranes are of the order of 40 to 80 millivolts it can be seen that the effect of pressure is relatively small. The sign is as expected, the excess pressure in phase α being counterbalanced by this phase becoming less positive and so repelling the positive sodium ions less. For anions the sign would of course be reversed, since $z = -1$.

Electrodes

The Donnan equilibrium serves as a very suitable introduction to the theory of electrodes. The physiologist meets with these in three main connections; firstly in the measurement of pH, secondly in the measurement of oxidation-reduction potentials, and thirdly, in the form of what are called salt bridges in the measurement of the electrical potentials of phases. The problem that must be discussed is how it comes about that inserting different electrodes into the same solution enables quite different measurements to be made upon it.

We start with the one of the fundamental equations derived earlier for the transport equilibrium of an ion (cf. equation (12.13), where it is in a slightly different form):

$$RT \ln\left(\frac{a_i^\alpha}{a_i^\beta}\right) = -z_i \mathscr{F} (\psi^\alpha - \psi^\beta). \tag{12.40}$$

This equation might be regarded as expressing the equivalence between a concentration ratio and an opposing difference of electrical potential, the one 'balancing' the other. It has also been seen in connection with non-

---

[1] The effect sought is of course measured by $\left(\dfrac{\partial \psi}{\partial P}\right)_{T, \mu_i, m_i}$ which can be obtained directly from (12.37) as $-\bar{V}_i/z_i\mathscr{F}$.

electrolytes (equation (11.17)) and ions (equation (12.39)) how a pressure difference can similarly offset the effect of either a concentration gradient or a membrane potential, and reduce a transport system to equilibrium. The fact that different physical 'forces', such as electrical potentials and pressure differences, can be regarded as equivalent to each other and to a concentration ratio introduces the possibility of using one of them to measure another, and this is particularly important in the case of the difference in electrical potential, which can be very readily and accurately determined.

We now return to a consideration of equation (12.40). This relates the properties $(a_i^\alpha, \psi^\alpha)$ of one phase to those of a second phase in equilibrium with it. Suppose we have two aqueous ionic phases; one (α) is an unknown solution, and the other (β) a solution of chosen composition. If these solutions are placed in contact through a suitable membrane they will exchange various ions and perhaps other substances until they achieve equilibrium, and in the process an electrical potential difference will generally result. The nature of the equilibrium will depend on the nature of the membrane; for instance, if the membrane is chosen to be permeable to *every* chemical species present then equilibrium will entail the two phases becoming identical in composition, with zero electrical potential difference. Suppose however that a membrane is chosen that is permeable *to only one ionic species*, i. Then at equilibrium equation (12.40) will hold for this species alone; and most importantly, an electrical measurement ($\psi^\alpha - \psi^\beta$) will enable the quantity a_i^α to be derived, since a_i^β is already known. Thus the problem of measuring the activity of i in the unknown solution is solved if

(1) a solution of known composition can be brought into contact with it through a membrane permeable only to i; and

(2) if the resulting electrical potential difference can be measured.

How the potential difference is measured will be dealt with in a moment, but firstly one or two other remarks must be made.

Suppose the membrane is permeable to a second constituent ion, j. How does this affect the measurement? The answer is that it upsets it completely, since both ions i and j are now free to enter or leave the phases *in chemically significant amounts*, and they will do so in co-ordination[1] until their relative concentrations satisfy the general relation (cf. equation (12.14))

$$\frac{a_i^\alpha}{a_i^\beta} = \frac{a_j^\alpha}{a_j^\beta} \tag{12.41}$$

modified if necessary for valencies other than unity. The electrical potential measured will of course still be related to these fractions by equation (12.40),

---

[1] That is, i may enter while j leaves in equal amount, if these ions are of the same sign; if of opposite sign they move together, in the same direction.

but in the process of mutual readjustment of the two ions the reference phase β will have changed significantly in composition with respect to i (and j), and consequently the electrical measurement no longer suffices to give us a_i^α. It is therefore absolutely necessary for the membrane to be permeable only to i. When this condition is met, the electrical potential is set up by the movement of a chemically insignificant amount of this ion (see p. 174) and a_i^β remains known.

It is thus seen that, in principle, an electrode fashioned on these lines can be used to measure the activity of any ion present in a solution of unknown strength. In practice, however, the great difficulty is to find a suitable membrane; but in two common cases this has been found quite feasible. In the glass electrode the delicate glass envelope functions in effect as if it were permeable only to the very small hydrogen ion, so that this electrode can be used to measure pH. Then again, in the case of a noble metal like platinum the only mobile component of the solution which can pass the interface and enter the metal consists of bare electrons; consequently the platinum electrode measures oxidation-reduction potentials. The metal in fact functions rather like a Donnan system in which positive charges are structurally bound and a membrane is unnecessary. Naturally, equation (12.40) still applies.

Measurement of the Difference of Potential

The next problem is how to measure the electrical potential difference[1] between two solutions (Fig. 12.5), a difference which represents in electrical terms the ionic activity ratio we wish to ascertain. Suppose metal wires (say silver ones) are simply inserted into the two solutions; would these suffice to pick up the potentials? The answer is no; but before the reason for this is discussed it is worth remarking that metal elements of some sort are required because all the usual types of laboratory measuring instruments possess metal terminals to which connection must be made, though it is quite possible to devise less convenient apparatus which avoids this requirement.

With regard to the question of why a bare silver wire placed directly in a solution is not satisfactory for picking up its potential, it may first be remarked that when a wire is inserted into a solution a wide variety of things may happen, many of them (perhaps all) to such a minute extent that they pass quite unnoticed. For instance, the wire goes into solution to an exceedingly small extent, and among the particles dissolved from its surface are silver atoms, silver ions, and bare electrons. Again, ions present in the solution are absorbed on to the silver. Now many of these changes are utterly insignificant chemically, but from the point of view of altering the potential of the wire they may be highly influential, as was discussed earlier in this

---

[1] In this chapter only *equilibrium* potentials are being considered, not mixing (diffusion potentials).

chapter. The trouble is that there is nothing *definite* about the whole situation; it is not inconceivable that a mere trace of contamination entering the system might send the potential of the wire soaring. It is this indefiniteness that a practical arrangement has to overcome.

Suppose a silver wire is placed in a solution in which, for the sake of argument, two well-defined processes can occur, both of them influencing it electrically. The first, we suppose, is the adsorption of an otherwise inert but surface-active dye cation; the second is the transfer of electrons from an oxidising-reducing mixture of ferric and ferrous salts. Were the dye alone

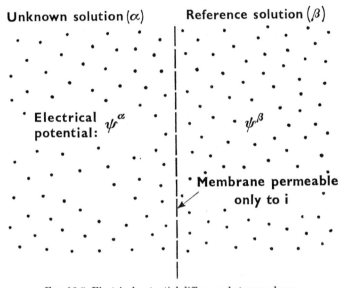

Fig. 12.5. Electrical potential difference between phases.

present it can be imagined that the silver wire would adsorb it to equilibrium and in the process assume a potential ψ' depending on the cation concentration. Were the mixture of iron salts alone present the silver wire would function as a noble metal and pick up the redox potential ψ''. What happens when the two systems are present together? The answer is that the silver wire assumes a potential ψ somewhere between ψ' and ψ'', and this comes about as follows.

Before the final potential ψ has been established the two systems can be imagined pulling against each other in a sort of tug-of-war. The dye system is attempting to establish the potential, ψ', and the redox system the potential ψ''. The redox system contributes an electron to the wire (to speak in concrete terms); the dye system responds by adding a positive ions to cancel the effect and pull the potential towards the value it 'wants'. The redox sytem then contributes another electron, and a further dye ion is absorbed, and so the process continues. The final outcome depends on which side can keep the process up longest; if the dye is very dilute and the redox salts are in fair concentration there will obviously come a time when the power of the dye system to donate ions will be exhausted, while the redox system has suffered very little change. The result will be that the value of ψ moves over very nearly (but not quite) to that of ψ''. On the other hand, if the dye is concentrated and the redox system of small chemical capacity then ψ will be nearly (though again not quite) equal to ψ'.

The simple situation sketched above brings out this important point: where a number of processes are jointly concerned in defining the potential assumed by a metal wire in a solution, that process is dominant which has the greatest 'staying power' or chemical capacity. If it has a great enough capacity it can 'swamp' the others and virtually dictate the potential alone. The trouble with the simple silver wire electrode is that the dominant process affecting it electrically is not known; and even if it were, there would still be insufficient knowledge of the conditions (such as the concentration of the dissolved reactants) under which it was taking place. A silver or some other type of wire must be used; but to overcome the above difficulties this is inserted into an auxiliary solution of known and suitable composition, and this solution is then connected electrically, by a method to be discussed later, with the unknown solution. The combination of metal wire and suitable solution constitutes what is called a reversible electrode.

REVERSIBLE ELECTRODES

Even a bare wire may reach equilibrium when immersed, and the passage of a minute current in either direction between wire and solution will then represent a thermodynamically reversible process; but what is implied by a reversible electrode is rather more than mere equilibrium. It may be said to involve firstly, a *single well-defined process* taking place under known conditions at the metal surface on the passage of a current; and secondly, the possibility of a current passing in either direction *for an appreciable time* without altering the state of affairs at the interface qualitatively, that is without any sort of polarization occurring[1]. Now the metal wire can be looked upon as an inexhaustible reservoir for two things: electrons, and its own species of metallic ion. The second requirement therefore means that the 'electrode process' must involve the transfer of either of these two entities between the wire and the solution. In the case of electrons the wire responds, as was seen, to the oxidation-reduction potential of the solution; and if this

[1] This means that the chosen process can dominate the situation and 'swamp' any uninvited processes in the sense just discussed.

can be made well-defined and of reasonably large capacity this is quite a satisfactory arrangement. In fact, the glass electrode is sometimes made with the interior phase consisting of buffered quinhydrone into which a bare platinum wire dips.

It is more common, however, to make the metal ions the entities which pass between the wire and solution; the electrode is made 'reversible to' silver ions, to take the chosen example. This involves dipping the silver wire into a solution containing silver ions. In order to make its electrical state quite definite the silver ions must have a fixed concentration, and this in turn suggests using a saturated solution of a silver salt. For reasons which will be touched on later a sparingly soluble salt like AgCl is preferable to a very soluble one like $AgNO_3$ and so this leads logically to the design of a silver-silver chloride electrode. Figure 12.6 which shows how the principle is applied to the case of a typical glass electrode, illustrates these points. The silver chloride is formed in direct contact with the silver wire, where it is naturally most effective. The 0·1N acid (acting as the reference phase β) maintains the internal pH constant (a strong acid acts as a buffer) and also provides a supply of chloride ions to partner new silver ions formed when the current is in the direction to take hydrogen ions outward through the glass membrane. Very similar remarks to the above apply to another very common arrangement, the mercury-calomel electrode.

The Salt Bridge

The final problem remains of connecting the unknown solution to be measured, and any reference solution against which it is being set, to the solutions into which the metal wires dip. (Incidentally, the reference solution may itself be one of the latter, as in Fig. 12.6.) This is the problem of forming liquid junctions in such a way that the unavoidable differences of potential they introduce will be least disturbing. Inevitably when two aqueous electrolyte solutions of differing composition come into free contact continuous interdiffusion takes place, and owing to the different rates at which ions diffuse this sets up diffusion potentials, as was remarked at the beginning of the chapter. Unfortunately these are not steady, but change with time as the concentration differences disappear; the best that can be done therefore is to take what steps are possible to reduce them in magnitude, and to cut down their inevitable fluctuations. Two simple expedients usually prove to be fairly adequate. The first is to make connections with solutions of a salt (like potassium chloride) whose ions move at almost identical rates, and therefore set up very small diffusion potentials; and the second is to work at fairly high concentrations of this salt (often at saturation), since it can be shown that this has the effect of 'swamping' the influence of other ions native to the unknown solution we are measuring. Incidentally, this is one reason why only sparingly soluble salts like silver and mercurous chlorides are used in reference electrodes; they leave the potassium chloride free to dominate the inevitable liquid junction.

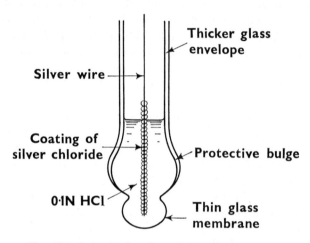

Fig. 12.6. A simple glass electrode for pH measurement.

As a practical convenience salt bridges of saturated potassium chloride are often made up in a medium of dilute agar instead of water, but the agar has no theoretical significance.

Arrangements for pH Measurement

The points that have been discussed can finally be embodied in a typical arrangement for the measurement of pH using the glass electrode (Fig. 12.7). The system can be described by writing it out as a formal scheme in the following way:

Cu	Ag	0·1N HCl saturated with AgCl	Glass membrane	Solution of unknown pH	Saturated KCl (salt bridge)	Saturated solution of Hg_2Cl_2 and KCl	Hg	Pt	Cu

where the vertical lines represent the boundaries between different phases, and the broken ones are liquid-liquid junctions. The copper wires to the measuring instrument have been included for completeness.

Referring to the previous discussion the 'reference solution' (β) with which the pH of the unknown is compared is the 0·1 N hydrochloric acid. The 'salt bridge' is formed by the lower part of the standard electrode; and the solution of fixed composition with which the mercury globule makes contact is saturated with mercurous and potassium chlorides. There are only two[1] undefined potentials developed in the circuit, and these are the liquid junction potential in the vicinity of the asbestos wick, and the all-important potential across the glass membrane. The former of course is small for the reasons already given, and in many practical designs is given increased constancy by ensuring a slow gravitational flow of saturated potassium chloride downward through

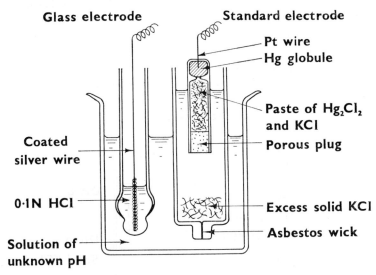

FIG. 12.7. Typical arrangement for pH measurement.

the wick, which keeps the concentration gradient at the junction fairly definite in form. It only remains therefore to find a link between the instrumental reading and the pH of the unknown solution, and this is done by replacing the latter with a buffer of known acidity. This done, the electrometer readings can be converted directly to pH. The relationship is linear, since both pH and electrical potential depend logarithmically on concentration.

OXIDATION-REDUCTION POTENTIALS

If the redox potential of the same solution is to be measured all that is necessary is to replace the glass electrode with a simple plate composed of one of the noble metals, say platinum. For such a metal no chemical reactions of significant extent are likely; consequently its electrical state can only be affected by its losing electrons to, or gaining electrons from, the

---

[1] There is a third at the porous plug, but the small solubility of mercurous chloride renders this negligible.

solution. This means the instrumental readings will reflect the oxidation-reduction potential of the latter, always assuming that the chemical capacity of the redox systems is adequate to swamp minute parasitic effects (such as the adsorption of ions) which may seek to assert themselves.

POTENTIAL OF A PHASE

The question of how to measure the potential of a liquid phase, or rather the difference of potential between two such phases in contact, has implicitly already been dealt with. The problem arises quite frequently in physiological work; one wishes for instance to measure the difference in potential between the vacuole of a large cell (such as an internodal cell of *Nitella*) and the external medium, or the potential across the membrane of a squid axon or a muscle fibre. In such cases each of the aqueous phases concerned is connected

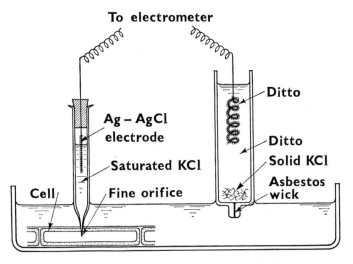

FIG. 12.8. Measurement of membrane potential of a plant cell.

by a salt bridge to standard solutions in which are immersed silver-silver chloride or other suitable reversible electrodes. Such salt bridges usually contain saturated potassium chloride, and the open end for insertion into the cell interior may be extremely minute, perhaps with an orifice of no more than half a micron in diameter. The combination of the salt bridge and the reversible silver-silver chloride electrode forms a standard electrode very similar to the one illustrated in Fig. 12.7. A typical circuit for measuring the membrane potential of a *Nitella* cell is shown above (Fig. 12.8). The two electrodes are identical except that one may be coarser in its construction.

Action of the Glass Electrode

This chapter will be concluded with a brief description of the action of the glass electrode, the particular form chosen being that in which a silver wire coated with silver chloride is immersed in 0·1N hydrochloric acid.

The fundamental process taking place at the metal-solution interface is the balanced exchange of silver ions:

$$\mathrm{Ag^+ \ (metal \ wire) \rightleftharpoons Ag^+ \ (dissolved)}.$$

At equilibrium therefore

$$\mu_{\mathrm{Ag^+}} \ (\text{wire}) = \mu_{\mathrm{Ag^+}} \ (\text{solution}). \tag{12.42}$$

When the glass electrode is placed in a solution (α) of pH greater than that of the 0·1N acid (β) inside it, hydrogen ions will pass outward across the glass membrane. This loss of positive charge will lower the electrical potential (ψ^β) of the acid, and so the electrochemical potential ($\mu^\beta_{\mathrm{H^+}}$) of its hydrogen ions. This will be accomplished with an entirely insignificant change in their actual concentration, a fundamental point we noted earlier. The process will come to a stop when the reduction in ψ^β is adequate to compensate for the concentration difference in $\mathrm{H^+}$ on the two sides of the glass (cf. (12.40)), so that

$$\mu^\alpha_{\mathrm{H^+}} = \mu^\beta_{\mathrm{H^+}}. \tag{12.43}$$

Meanwhile however, the reduction in ψ^β has also lowered the electrochemical potential of the silver ions in solution, and as a consequence the equilibrium described by equation (12.42) has been disturbed. In teleological language, the wire redresses the balance by throwing off positive silver ions into the solution. This has three compensating effects: it lowers the electrical potential (ψ^{wire}) of the silver wire (and thus $\mu^{\mathrm{wire}}_{\mathrm{Ag^+}}$); it raises the potential (ψ^β) of the solution; and it increases the concentration ($m^\beta_{\mathrm{Ag^+}}$) of the dissolved silver ions. These latter two effects would both raise $\mu^\beta_{\mathrm{Ag^+}}$, the term on the right hand side of equation (12.42), and so help to re-establish this equality; but in fact they are not allowed to, since the extra silver ions dissolved off the wire are at once precipitated[1] as chloride, the solution being already saturated with this salt. The result is that the whole task of restoring the balance represented by equation (12.42) has to be performed by the first effect, that is the fall in the electrical potential of the silver wire. As a consequence of this the potential difference appearing across the glass membrane is transferred *in toto* to the silver wire. Were it not that the concentration of silver

---

[1] This precipitation may be regarded as drawing to the vicinity of the wire the chloride ions whose hydrogen ion partners have crossed the membrane. This cancels the second effect.

ions is maintained at a constant value this would not have been so. When the external solution is *more* than 0·1N acid the sequence of events is the reverse of the one described, silver ions being deposited on the wire from the solution and some of the solid silver chloride being dissolved to replenish them. In this process it might be thought that the chloride ion concentration must change a little, and so also the silver ion concentration (they are linked through the solubility product of the insoluble salt); but a moment's reflection will show that the actual number of chloride ions involved is about the same as that of the hydrogen ions penetrating the membrane. Since the chloride ions are very abundant the proportional change in their concentration is therefore excessively minute, in fact entirely negligible. The same therefore applies to the silver ions.

CHAPTER 13

The Thermodynamics of Water Relations

'Now then, Roo, dear,' she said, as she took Piglet out of her pocket.
'Bed-time.'
'Aha!' said Piglet, as well as he could.
'Bath first,' said Kanga in a cheerful voice. 'I am not at all sure that it wouldn't be a good idea to have a *cold* bath this evening. Would you like that, Roo, dear?'

WINNIE-THE-POOH
A. A. Milne

The subject of the water relations of cells and tissues is one of perennial interest to both plant and animal physiologists, and it is likely to remain so for a long time. Starting with an early recognition of the importance of osmotic forces in water movement in tissues (Dutrochet[1], 1827), a great advance was made when it was recognized that turgor pressure in plant cells, maintained by the elastic properties of the cell walls, was also a factor of fundamental importance (Ursprung and Blum[2], 1916; Thoday[3], 1918). What is still an interesting and largely unanswered question is the extent to which water movement is also brought about by active agencies; that is, agencies which superficially accomplish the same sort of water transport as passive ones, but which can only maintain the redistribution of water they have brought about by the continuous dissipation of energy. The further consideration of active mechanisms is best deferred to the last chapter, when the Thermodynamics of Irreversible Processes will be discussed. Meanwhile there is something to be said about the interaction of ordinary passive forces in water movement.

THE FUNDAMENTAL RELATION

Passive forces of course lead to true thermodynamic equilibrium; and from what was said in a previous chapter (Chapter 11) the reader will be prepared to accept the statement that the fundamental criterion for water equilibrium is equality of its chemical potential (μ_w) between all phases. Further, the direction in which water will move between two phases α and β in contact is given by the sign of $(\mu_w^\alpha - \mu_w^\beta)$. This of course assumes the absence of active mechanisms, which in any case imply a denial of true equilibrium.

Thus for a cell in equilibrium

$$\mu_w^{\text{external milieu}} = \mu_w^{\text{cell wall}} = \mu_w^{\text{cytoplasm}} = \mu_w^{\text{vacuole}}, \qquad (13.1)$$

where, to name two of these phases, the cell wall may be holding its water

---

[1] Dutrochet, R. (1827). *Ann. Chim. (Phys.)* **35**, 393.
[2] Ursprung, A. and Blum, G. (1916). *Ber. dtsch. bot. Ges.* **34**, 88.
[3] Thoday, D. (1918). *New Phytol.* **17**, 108.

by an imbibitional mechanism while the vacuole does so by an osmotic one, the question of how the water is held making no difference to the relationship (13.1).

Before equation (13.1) can be developed a general expression must be obtained for μ_w similar to equation (12.37) of the previous chapter, but with two modifications. In the first place, since the water molecule has no net charge $z = 0$ and the term involving ψ disappears. In the second place, the term, $RT \ln \gamma_i m_i$ will need examination, since it was developed with special reference to *solutes*, and there is clearly no meaning in reading m_i as the molality of the *water*. As a matter of fact it is a very simple matter to rewrite this term. Suppose in the phase α the osmotic potential of the water is Π^α and likewise Π^β in phase β. Then the transfer of dn_w moles of water between the two phases, involving a transfer of volume $\bar{V}_w dn_w$, can be accomplished as follows. A pressure equal to Π^α (or infinitesimally greater) is applied to the first phase, and this volume of water is squeezed out of it reversibly through a semi-permeable membrane, the work done being $\Pi^\alpha \bar{V}_w^\alpha dn_w$. This water is then allowed to enter the second phase, similarly against the pressure Π^β (or one infinitesimally smaller), the work received being $\Pi^\beta \bar{V}_w^\beta dn_w$. The net result is that the transfer has been accomplished reversibly and the work obtained is $(\Pi^\beta \bar{V}_w^\beta - \Pi^\alpha \bar{V}_w^\alpha) dn_w$. Equating this to the formula in terms of chemical potentials:

$$\text{maximum work} = (\mu_w^\alpha - \mu_w^\beta) dn_w$$

gives

$$\mu_w^\alpha - \mu_w^\beta = -(\Pi^\alpha \bar{V}_w^\alpha - \Pi^\beta \bar{V}_w^\beta), \tag{13.2}$$

where the negative sign enters because the higher the osmotic potential of the phase, the lower the chemical potential of its water and vice versa. It is not difficult to appreciate therefore that the terms $RT \ln \gamma_i m_i$ of equation (12.37) must be replaced by $-\Pi \bar{V}_w$ and the equation we require becomes

$$\mu_w = \mu_w^0 - \Pi \bar{V}_w + P \bar{V}_w. \tag{13.3}$$

It is easy to see the meaning of μ_w^0. It is the potential of the water when $\Pi = 0$ (i.e. when it is pure) and when the external pressure (P) on it is also zero (liquid water would of course then be highly metastable, but this is incidental). Thus when equation (13.3) is applied to two phases, the μ_w^0s will cancel. Incidentally, it is obviously appropriate to drop the old superscript $^\ominus$ and to use instead ^{0}.

Before proceeding with the argument two interesting points will be discussed. The reader may have noticed that two values have been used for the partial molar volume of water ($\bar{V}_w^\alpha$ and $\bar{V}_w^\beta$) since one phase will not, in general, decrease in volume by exactly the same amount as the other increases. However in biological systems, except for very dry colloids, $\bar{V}_w$ can usually be taken as constant and equal to the molar volume of pure water (i.e. 18·0 millilitres).

Next suppose for simplicity that there is only one solute (i) present. We can write approximately

$$\Pi = RTc_i = RT\sigma m_i, \tag{13.4}$$

where σ has the dimensions of density and is nearly equal to unity. Equation (13.3) thus becomes

$$\mu_w = \mu_w^0 - RT\sigma m_i \bar{V}_w + P\bar{V}_w. \tag{13.5}$$

For comparison, we may write for the solute

$$\mu_i = \mu_i^\ominus + RT \ln m_i + P\bar{V}_i, \tag{13.6}$$

where the activity γ_i has been omitted on the assumption that we have an ideal dilute solution.

These two equations show that if the molality, m_i of the solute is changed the chemical potentials of both the water and the solute alter. This is naturally to be expected; but these equations also enable the relation between the two changes, $d\mu_w$ and $d\mu_i$ to be found. If they are differentiated with respect to m_i (at constant T,P) then we have

$$d\mu_w = -RT\sigma \bar{V}_w dm_i \tag{13.7}$$

$$d\mu_i = \frac{RT}{m_i} dm_i \tag{13.8}$$

and for equal values of dm_i in the two cases, eliminating dm_i by division gives

$$\frac{d\mu_w}{d\mu_i} = -\sigma m_i \bar{V}_w. \tag{13.9}$$

This shows that the two chemical potentials change in opposite directions. Further, since σm_i is the volume concentration (see equation (13.4)) and the solution is very dilute, the quantity $\sigma m_i \bar{V}_w$ is simply $\frac{n_i}{n_w}$, where n_i and n_w are the number of moles of solute and water present respectively. For a dilute solution, the fraction $\frac{n_i}{n_w}$ is small; consequently the change in the chemical potential of the water is small compared with the corresponding change for the solute. Equation (13.9) can be rewritten $\left(\text{putting } \sigma m_i \bar{V}_w = \frac{n_i}{n_w}\right)$

$$n_w d\mu_w + n_i d\mu_i = 0 \quad \text{(constant } T,P\text{)} \tag{13.10}$$

and although it has been derived rather clumsily and only for an ideal dilute solution, this is in fact a particular case of an *exact* and quite general equation called the Gibbs-Duhem equation. It applies to any number of components, not merely to two, and to mixtures as well as to solutions.

The Suction Potential of Cells and Tissues

We now continue with the main argument. For two phases in equilibrium with respect to water movement we can write, using (13.3),

$$\mu_w^0 - \Pi^\alpha \overline{V}_w^\alpha + P^\alpha \overline{V}_w^\alpha = \mu_w^0 - \Pi^\beta \overline{V}_w^\beta + P^\beta \overline{V}_w^\beta$$

or, cancelling μ_w^0 and rearranging,

$$(\Pi^\alpha - P^\alpha)\overline{V}_w^\alpha = (\Pi^\beta - P^\beta)\overline{V}_w^\beta. \tag{13.11}$$

Provided the partial molar volume of water can be considered as a constant, the criterion for equilibrium between phases is therefore equality of the quantity $(\Pi - P)$, and the plant physiologist will recognize that this is what he has long been used to calling the suction potential, diffusion-pressure deficit, suction pressure and a variety of other names. It is linked with the chemical potential of the water by the equation

$$\Pi - P = \left(\frac{\mu_w^0 - \mu_w}{\overline{V}_w}\right) \tag{13.12}$$

(cf. equation (13.3)) where μ_w^0 may be taken to refer to pure water at whatever pressure[1] serves as the zero for measuring P, conveniently atmospheric; $\overline{V}_w$ is virtually a constant, the molar volume of water (18·0 millilitres). It need hardly be added that in terms of classical plant physiology and with reference to the vacuole, Π is the osmotic potential of the cell sap and P the turgor pressure, or a little less satisfactorily (since it may omit reference to the pressure of neighbouring cells) the wall pressure.

The Question of Name

The question of what name to give to the quantity $(\Pi - P)$ is one which has divided plant physiologists for a long time. The three best are probably suction potential, diffusion-pressure deficit, and water potential. Of these the last is rather close to the well-established 'chemical potential of water' without being by any means the same thing; while the second has the disadvantage that it is negative in emphasis and seems to imply that water movement is a matter of diffusion only, forgetting that there is an energy term in the expression for free energy as well as an entropic one[2]. Con-

---

[1] This practical indifference to the datum level for P is a consequence of the virtual incompressibility of watery solutions.

[2] A moment's thought will convince the reader that diffusion (and hence osmosis) is linked with the tendency of *entropy* to increase, flow under pressure with the tendency of *energy* to decrease. See further below.

sequently the first term is the one preferred here. It is positive in emphasis, quite correctly suggestive of a *potential suction* (realized for instance in the transpirational pull), and in addition has a fair degree of historical priority. However it is hardly a difficult matter to accept the others as alternatives.

Needless to say, equality of suction potential ceases to be the criterion for zero net water movement when active mechanisms are present, and this includes all cases where equality of temperature does not hold. In fact, quite small differences of temperature can completely overshadow large differences of suction potential; one degree centigrade is equivalent to over eighty atmospheres, in the sense that water will distil *from* an open solution of eighty atmospheres osmotic potential *to* pure water maintained at only one degree centigrade below it. Thus the comparison of suction potentials as a means of determining the direction of water movement is rarely satisfactory unless the tissues concerned are close together; between the roots and leaves of a tall plant for instance it will certainly rarely be so. The problem of water movement in the presence of temperature gradients is one which can best be dealt with by the methods of the Thermodynamics of Irreversible Processes, and this newer development is discussed in the final chapter.

Three Examples of Water Movement

It will be appreciated that under conditions of uniform temperature, the chemical potential, which uniquely determines the direction of movement of a substance, is composed of terms which reflect the influence of both energy and entropy, just as in the case of plain free energy. To illustrate this for water we can write (since $G = H - TS$, and differentiating with respect to n_w)

$$\mu_w = \frac{\partial G}{\partial n_w} = \frac{\partial H}{\partial n_w} - T\frac{\partial S}{\partial n_w}$$

$$= \overline{H}_w - T\overline{S}_w \qquad (13.13)$$

where $\overline{H}_w$ and $T\overline{S}_w$ are respectively the energetic and entropic terms. Or again, referring to the general formula (12.37) for a solute,

$$\mu_i = \mu_i^\ominus + RT \ln \gamma_i m_i + z_i \mathscr{F}\psi + P\overline{V}_i \qquad (13.14)$$

it will be seen that $z_i\mathscr{F}\psi$ and $P\overline{V}_i$ are essentially energetic components and $RT \ln \gamma_i m_i$ is in the main an entropic one. In other words, when the chemical potential difference which is responsible for the movement of the substance i arises from the electrical and pressure terms, the movement of i is being promoted by actual forces (which could in principle be measured on a spring balance), and this implies the involvement of energy; but when it is due to the concentration term the net movement of i results essentially from the randomizing tendency of molecular thermal movements, i.e. it is entropic[1].

Consider now three typical cases of (isothermal) water transport in plants, all of which, it should be emphasized, are describable in terms of chemical potential or suction potential gradients. The first is the mass flow of water along the xylem in response to a transpirational pull. Here the suction potential gradient (between two points within the xylem) arises almost entirely in connection with the pressure term P, there being virtually no contribution from the concentration or osmotic term Π. If we call the suction potential $S.P.$ then in general

$$S.P. = \Pi - P \qquad (13.15)$$

or

$$\Delta(S.P.) = \Delta\Pi - \Delta P.$$

But for xylem flow, since $\Delta\Pi$ is probably minute over reasonable distances,

$$\Delta(S.P.) = -\Delta P. \qquad (13.16)$$

The process is thus an *energetic* one, and has nothing to do with changes in the entropy of the sap. Its spontaneity arises from the fact that it lowers the energy of the system, and thus also the free energy.

The second example concerns the uptake of water by a plasmolysed protoplast. Since the protoplast has very little mechanical strength the turgor pressure is almost nil. The suction potential differential therefore arises almost completely from the osmotic term:

$$\Delta(S.P.) = \Delta\Pi \qquad (13.17)$$

and the movement of water owes its spontaneity to the fact that it raises the entropy of the system[2]. Like all movements of simple diffusion it is in the direction of a more probable distribution.

The third example is the flow of water across the cortical cells of the root. Here adjacent cells may have appreciably different turgor pressures and likewise osmotic potentials, the suction potential difference between two cells being made up of both components:

$$\Delta(S.P.) = \Delta\Pi - \Delta P. \qquad (13.18)$$

The movement therefore is neither purely energetic nor purely entropic, but a combination of the two.

These three examples illustrate how the interplay of both energy and entropy enter into the movement towards equilibrium, a point that was noticed earlier in Chapter 7. Diffusion being a process associated essentially

---

[1] This is not to deny that energy *may* also have a part here; there may be a heat of dilution involved, as in the diffusion of strong sulphuric acid into water.

[2] This again lowers the free energy, the appropriate criterion in isothermal situations.

with the tendency of entropy to increase (and having no *necessary* connection with macroscopic energy changes), these considerations emphasize again the inadequacy of the term 'diffusion-pressure deficit' for the quantity that has been called the suction potential.

One further point might be noted before leaving this subject; it concerns the uptake of water by imbibition. This is best regarded as an energetic process, and where imbibing colloids are present an extra term can be added to equation (13.3) to allow for their influence on the chemical potential of the water. The matter need not, however, be pursued further.

Connection between Vapour Pressure and Suction Potential

In concluding this chapter we shall allude briefly to the connection between suction potential and equilibrium vapour pressure. The relationship between osmotic potential and vapour pressure was discussed from an elementary point of view in an earlier chapter (Chapter 7); the present treatment will be rather more general.

Consider two phases α and β (or positions in a continuous medium with non-uniform properties) where the chemical potential of the water is μ_w^α and μ_w^β. Then with the previous notation we can write

$$S.P.^\alpha - S.P.^\beta = \left(\frac{\mu_w^\beta - \mu_w^\alpha}{V_w}\right) = (\Pi^\alpha - P^\alpha) - (\Pi^\beta - P^\beta). \tag{13.19}$$

This is the general equation, and naturally it is taken for granted (unless otherwise stated) that the temperature is uniform.

Suppose that the aqueous vapour pressures in equilibrium with the phases α and β are p^α and p^β respectively. Then the chemical potentials of this vapour are equal[1] to μ_w^α and μ_w^β. However, considering the vapour, the difference between these two potentials can easily be derived from the gas laws, for the work done in the isothermal expansion of dn_w moles of vapour from p^α to p^β is simply $RT \ln \frac{p^\alpha}{p^\beta} dn_w$ (Chapter 5). Thus (from the fundamental definition of μ)

$$(\mu_w^\alpha - \mu_w^\beta)dn_w = RT \ln \frac{p^\alpha}{p^\beta} dn_w$$

or

$$\mu_w^\beta - \mu_w^\alpha = RT \ln \frac{p^\beta}{p^\alpha}, \tag{13.20}$$

---

[1] This follows of course from the condition for equilibrium of vapour and liquid phases. Strictly, p depends on P (which differs between phases), but the effect is small and is overlooked in what follows. For $\Delta P = 1$ atmosphere the proportional change in p is about 0·07% for water at 25°C.

inverting the fraction to obtain the μ's in the required order (since the suction potential varies in the opposite direction to the chemical potential).

Combining equations (13.19) and (13.20) we have

$$S.P.^{\alpha} - S.P.^{\beta} = \frac{RT}{\overline{V}_w} \ln \frac{p^{\beta}}{p^{\alpha}}, \qquad (13.21)$$

and if phase β happens to be pure water at the zero of pressure (usually atmospheric), then $S.P.^{\beta}$ is zero, p^{β} can be written p^0 (where the superscript refers to saturation) and so

$$S.P. = \frac{RT}{\overline{V}_w} \ln \frac{p^0}{p} \qquad (13.22)$$

dropping the superscript α as this is no longer required. It is important to realize that this relation applies to the suction potential quite generally, but not to the osmotic potential. However, if the pressure on phase α happens to be that chosen as zero (this means the pressure, usually atmospheric, at which pure water has the measured saturation vapour pressure p^0) then the suction potential and the osmotic potential are equal (see equation (13.15)), and the relation for osmotic potential derived in Chapter 7 is obtained

$$\Pi = \frac{RT}{\overline{V}_w} \ln \frac{p^0}{p}. \qquad (13.23)$$

This is a valuable formula, and enables the osmotic potential of solutions like strong salts or acids to be calculated, cases which might otherwise be difficult to handle. The value of $\overline{V}_w$ should ideally be taken as the mean of that for pure water at atmospheric pressure, and that for the solution under its osmotic pressure; but in ordinary cases it is adequate to take it as simply the molar volume of water.

The general equation (13.22) indicates perhaps the best way of all for measuring the suction potential of tissues or organs. In principle the organ, such as a leaf, has merely to be confined in a closed space and the vapour pressure over it measured; but as ten atmospheres suction potential means a drop of only about three-quarters per cent in the relative humidity[1], and as vapour pressure is very sensitive to temperature, highly accurate temperature control is needed and special techniques are required suitable for carrying out measurements so close to saturation.

Postscript

The reader may have noticed in connection with equation (13.11) a fact not usually appreciated: that it is not quite true to say that equality of

[1] This can easily be verified from equation (13.22).

13. THE THERMODYNAMICS OF WATER RELATIONS

suction potential is the criterion for water equilibrium. This is not usually a point of any practical importance, but it is theoretically interesting. It is a simple corollary of Lewis's statement of the Second Law that when equilibrium obtains the transfer of an infinitesimal amount of water between two phases yields (or requires) no net work, since as much work has to be expended in 'squeezing out' the water from one phase (against its suction potential) as is restored when the other phase, exerting its suction potential, absorbs it. Thus at equilibrium it is not fundamentally two opposing suction potentials which are in exact balance, but two (virtual) quantities of work. This accounts for the appearance of the partial molar volumes in equation (13.11); in fact this equation may be read as a statement of the equality of the two amounts of work when one mole of water is transferred from one phase to the other.

The same remarks apply, naturally, to the simple case of two isotonic solutions. They are not in general in exact mutual equilibrium. In fact they are only so when under their actual common osmotic pressure[1]. Practically, however, this is a very small point.

NUMERICAL VALUE OF THE PARTIAL MOLAR VOLUME OF WATER

As an example of the values to be expected for $\bar{V}_w$ and how they can be derived from published data two typical cases are calculated, potassium chloride solution at a concentration of 21% w/w (roughly 3 molar) and sucrose solution of 30% w/w (roughly 1 molar). The data in columns (1), (2) and (3) are taken from the Handbook of Physics and Chemistry[2], 13th edition; they refer to 20°C.

(1) Solution w/w	(2) Density kilogram per litre	(3) Solute kilogram per litre	(4) Total volume (l) per kilogram solute	(5) Total mass (kg) solute
KCl $\begin{cases} 20\% \\ 22\% \end{cases}$	1·1328 1·1474	0·2266 0·2524	4·413 3·962	5·000 4·545
	Increases		0·451	0·455

$$\text{Partial molar volume (20 to 22\%)} = \frac{0·451}{0·455} \times 18·00$$

$$= 17·84, \text{ say } 17·8 \text{ millilitres.}$$

---

[1] The question may arise why, since this pressure is the same for both (isotonic) solutions, they are not in equilibrium at any other common pressure. The answer involves their compressibility as liquids (see footnote one, p. 206).

[2] Chemical Rubber Publishing Co., Cleveland, Ohio.

	(1)	(2)	(3)	(4)	(5)
Sucrose	28%	1·1175	0·3129	3·196	3·571
	32%	1·1366	0·3637	2·750	3·125
			Increases	0·446	0·446

$$\text{Partial molar volume (28 to 32\%)} = \frac{0\cdot 446}{0\cdot 446} \times 18\cdot 00$$

$$= 18\cdot 0 \text{ millilitres.}$$

Note that the figures in column (4) are the reciprocals of those in (3); and that those in column (5) (apart from the decimal point) are the reciprocals of those in (1). Alternatively, the latter are the product of those in (2) and (4). The figure 18·00 appearing in the final calculation is the molar weight of water.

For comparison, the molar volume of water at 20°C (density = 0·99823) is 18·03 ml, and the partial molar volume in saturated sodium sulphate (24% anhydrous salt) is 17·6 ml.

CHAPTER 14

Photosynthesis, Thermodynamic Efficiency and ATP

> 'I'm planting a haycorn, Pooh, so that it can grow up into an oak-tree,' said Piglet. 'Christopher Robin says it will.'
> 'Well,' said Pooh, 'if I plant a honeycomb outside my house, then it will grow up into a beehive.'
> Piglet wasn't quite sure about this.
> 'Or a *piece* of a honeycomb,' said Pooh, 'so as not to waste too much. Only then I might only get a piece of a beehive, and it might be the wrong piece, where the bees were buzzing and not hunnying. Bother.'
>
> THE HOUSE AT POOH CORNER
>
> *A. A. Milne*

As one of the most important aspects of plant growth, photosynthesis and the efficiency of this process, will always be of great interest to the physiologist. In this chapter one or two facets of this problem will be considered, but these will be limited fairly drastically to thermodynamics and questions of mechanism will not be discussed.

EFFICIENCY

The notion of the efficiency with which a process involving energy transformation is carried out is one which comes up fairly frequently. However the word is not always employed in precisely the same sense, and this is apt to cause some confusion. Consider for instance an ordinary heat engine. In its simplest form such an engine takes in a quantity of heat Q_1 at a higher temperature T_1 and rejects a quantity Q_2 in its exhaust at a lower temperature T_2. The balance $W = Q_1 - Q_2$ is delivered as work. Now it might be said that the efficiency of the transformation of heat into work is W/Q_1 since usually Q_1 has to be paid for (and Q_2 rarely constitutes a refund) and all there is in return for it is W. This fraction (that is, W/Q_1) might therefore perhaps be called the *economic* or *technological* efficiency. However, the Second Law shows that even with the most idealized and perfect machinery the fraction W can only rise to the value $(T_1 - T_2)/T_1$. Thus it might be said that if it actually has this value the conversion reaches, on the basis of the Second Law, an efficiency of one hundred per cent. In general for less perfect engines and on the same basis, the *thermodynamic* efficiency might be measured by the ratio $W \left/ \left(\dfrac{T_1 - T_2}{T_1} \right) Q_1 \right.$.

Besides these two uses there are a number of others, such as the one embodied when speaking of the 'quantum efficiency' of a process. This is again rather different, but it is unlikely to cause confusion here.

The Criterion for Perfect Thermodynamic Efficiency

A process can be said to be carried out with perfect thermodynamic efficiency when it is conducted reversibly. This of course is always an ideal state of affairs, but it is important in that it provides a standard by which the actual efficiency of the process can be judged. By this standard it can be said that the thermodynamic efficiency is measured by the fraction

$$\eta = \frac{\text{useful result obtained (in work units)}}{\text{maximum thermodynamically possible}}$$

and the value for a perfect heat engine is thus unity, or one hundred per cent.

The problem to be solved is how, in the general situation, the 'maximum thermodynamically possible' can be found. It is fairly obvious that in order to do this further information must be obtained about the conditions under which the process is carried out. For instance, a very important set of conditions is that the reacting system is in contact with a reservoir which will provide it with all the heat it needs to keep its temperature constant; and which also has the ability to do all the necessary work on the system to keep its pressure constant. Granted these resources, the maximum work the system can do is represented by its free energy change, and the thermodynamic efficiency is measured by the fraction

$$\eta = \frac{\text{work done or free energy fixed in (for example) chemical form}}{\text{free energy decrease of primary reaction}}.$$

The condition[1] for this to be one hundred per cent can be expressed very simply in the form:

overall free energy change of system ΔG = zero;

$(\eta = 100\%;\ \text{constant}\ T,P)$

since what one chemical (or osmotic or elastic) system loses the other exactly gains.

This criterion is of course already well known to the reader; however, the present chapter is concerned with a more general one, and this is provided by the most general statement thermodynamics offers of the way in which spontaneous processes occur. As we saw in Chapter 8 this is that in an isolated system (another way of saying that everything affected by the change is included in the computation) all spontaneous processes involve an increase in entropy. In the limit, when the process is conducted reversibly, the entropy

---

[1] Where external work is delivered (such as raising weights) the condition stated must be modified to include this. The point is unimportant, however, in the present context.

increase is zero. At the risk of repetition it may be said in passing that this condition, and the free energy one, are not fundamentally two quite different criteria. Rather they are two different methods of applying the same criterion; one applicable to the general situation, and the other to a particular one, where in return for sacrificing generality (that is, accepting constancy of temperature and pressure) the free energy offers the compensating advantage of requiring only those changes occurring *inside* the system to be taken into account, and of ignoring those which accompany them in the surroundings. However, the entropy rule could still be applied if desired.

To return to the point at issue, the general condition sought which will define one hundred per cent efficiency is clearly

overall entropy change ΔS (system and surroundings) = zero;

($\eta = 100\%$; no restrictions on T, P, etc.).

Although in so doing we put the cart before the horse, historically speaking, it is instructive to pause for a moment to find the 'economic' efficiency of a perfect heat engine using this result. Since the engine works in a cycle its own entropy changes can be neglected; hence those of the surroundings only have to be considered. When the heat Q_1 is absorbed reversibly the loss of entropy of the source is, by definition, Q_1/T_1. Correspondingly, when the heat Q_2 is rejected the gain is Q_2/T_2. The overall increase in entropy of everything concerned is thus $\left(-\dfrac{Q_1}{T_1} + \dfrac{Q_2}{T_2}\right)$, and for perfect efficiency

$$\Delta S = -\frac{Q_1}{T_1} + \frac{Q_2}{T_2} = 0$$

that is,

$$\frac{Q_1}{Q_2} = \frac{T_1}{T_2}$$

and

$$\eta = W/Q_1 = (T_1 - T_2)/T_1 \tag{14.1}$$

since $W = Q_1 - Q_2$. Of course this is being wise after the event, for equation (14.1) was originally derived by Carnot for a perfect gas engine before entropy was even thought of; but it is instructive to see it as an illustration of calculating the maximum efficiency in a non-isothermal case.

The Problem of Photosynthesis

What happens in photosynthesis is that radiation out of equilibrium with the leaf is absorbed, and its energy used to perform work[1], this work being

---

[1] In Brønsted's wider use of the term (for full reference see footnote three, p. 13).

represented by the storage of chemical free energy. The chemical reactions involved take place in a system which may be considered as being at constant temperature and pressure. Since thermodynamics is not concerned with mechanism, the process may be split up for purposes of analysis in any suitably convenient way. Suppose this is done as follows, the stages being conducted with perfect reversibility.

Stage 1. Energy of radiation absorbed and used to do (mechanical) work, such as the coiling of a spring;

Stage 2. Stored work used to promote chemical syntheses in *isothermal* system open to atmospheric pressure.

Split up like this it can be seen[1] that the energy stored in the spring must be exactly equal to the overall free energy change as normally calculated for the reaction

$$6CO_2(g) + 6H_2O(l) \rightarrow C_6H_{12}O_6(aq.) + 6O_2(g)$$

and the problem thus reduces to that of *Stage 1;* with what economic efficiency can the energy of radiation be converted into work? Can it *all* be realized as work, or is it like the energy taken in by a heat engine, only part being convertible and the rest of necessity being rejected as heat? In answering this question use will have to be made of the criterion of zero entropy change; but before this can be done a digression is desirable to discuss some of the thermodynamic properties of radiation.

Thermodynamic Properties of Radiation

Just as a sample of gas can be contained in a vessel with *impervious* walls, so a sample of radiation can be regarded as imprisoned in a vessel with *perfectly reflecting* walls. If in such a vessel two small separated bodies are also enclosed, one with an ideally black surface of zero reflecting powers and the other with a surface of normal colour, experience shows that eventually the two bodies will attain equality of temperature. In this condition they will be exchanging thermal radiation; but whereas the black body absorbs all that falls on it the other absorbs only a fraction and reflects (or perhaps transmits) the rest. It follows that the black body must be a better *radiator* than the other in exactly the same ratio as it is a better *absorber*. In fact, the radiating powers of all bodies at that temperature must be in exactly the same ratio as their absorbing powers, or they could never attain temperature equality (Kirchhoff's Law). In particular, if a body is perfectly transparent to all wavelengths except those within a narrow band, then it will be quite incapable of emitting any radiations except those within this band. While no substance can be found which perfectly fits this specification (any more

---

[1] The significance of this analysis is that *Stage 2* is a simple *isothermal* one which can be handled in terms of free energy.

14. PHOTOSYNTHESIS, THERMODYNAMIC EFFICIENCY AND ATP

than a perfect gas can be found), the existence of examples which approximate to it for particular wavelengths can be used to justify the conviction that in postulating such a substance we are not supposing something essentially contrary to the laws of nature; consequently our conclusions will be valid ones.

Monochromatic Radiation

It will result in a simplification if discussion of photosynthesis is limited to monochromatic radiation, and where appropriate this will be taken as having a wavelength corresponding to the fluorescence maximum of chlorophyll in red light (6,800 Å), since this seems to be for the green plant, a very important wavelength. However, the thermodynamic principles are no different for other wavelengths. Further, when speaking of monochromatic parallel radiation it must be remembered that of physical necessity such radiation will occupy a finite band of the wavelength spectrum (say from λ to $\lambda+\delta\lambda$) and be dispersed through a finite solid angle[1] ($\delta\Omega$). In other words, parallel monochromatic radiation is never *exactly* parallel, nor of a single *exact* wavelength.

Suppose there are two small bodies both composed of a substance perfectly transparent to all parts of the spectrum except for the wavelength λ. These are introduced into a vessel with perfectly reflecting walls which is previously empty of both radiation and matter. (As a point of interest this could be done by conducting the operation inside a larger vessel maintained indefinitely near the absolute zero of temperature; and the two experimental bodies could be given suitable temperatures later by an internal delayed-action mechanism.) If the two bodies are initially at different temperatures they will naturally in time come to the same temperature, but they will do this solely by exchanging radiation of the wavelength λ. At final thermodynamic equilibrium the reflecting enclosure will contain two bodies at a temperature, T together with isotropic radiation of wavelength λ. It is therefore obvious that this radiation must be regarded as having the temperature T. Next, if a fairly large[2] black body having the same temperature T is introduced into the enclosure the equilibrium will be left undisturbed and the intensity of the wavelength, λ remain as before, although the cavity will now contain radiation of all wavelengths. This radiation will possess a continuous spectrum and a particular spectral distribution of energy, that of so-called 'black body radiation of temperature, T'. We may conclude therefore that the temperature of isotropic monochromatic radiation is equal to that of a black body whose own equilibrium radiation has the same intensity *at the same wavelength* (Fig. 14.1).

[1] See Appendix I for a definition of this mathematical concept.
[2] The black body is fairly large so that it can fill the space with its own radiation without losing temperature. If it were a mere 'speck' (see later) it could not do this.

These considerations enable us to see the logic firstly, of attributing a temperature to isotropic monochromatic radiation; and secondly, of relating this temperature to black-body radiation. This latter point is important, since Planck in 1900 discovered the formula[1] describing the distribution of energy over the electromagnetic spectrum for black body radiation; and this provides a means of calculating the temperature of isotropic monochromatic light given its intensity.

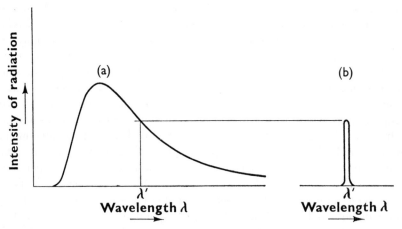

FIG. 14.1. Comparison of the intensity of monochromatic radiation from a selectively emitting body (b) with that at the same wavelength (λ') from a black body (a) at same temperature.

Further, however, a temperature can be attributed to *beamed* (that is, non-isotropic) light. Suppose a minute hole is made in the wall of the reflecting cavity. The monochromatic radiation will emerge uniformly over a solid angle of 2π (Fig. 14.2), and if the hole is suitably placed at the focus of a parabolic mirror the radiation will be converted into a parallel beam. This may be caught by an exactly similar arrangement facing it, brought to a focus and passed into a second reflecting cavity, where it will be in equilibrium with a similar object at the same temperature, T. Since the whole assemblage is in temperature equilibrium we must attribute to the monochromatic radiation everywhere the same temperature, whether it is isotropic (at A), in the form of a divergent beam (at B), or in parallel rays (at C). What is common at all these positions is not the luminous intensity (that is, the amount of radiant energy incident on a unit area per second) but the intensity *per unit solid angle*. This is particularly evident in the divergent pencil, B. The rays diverge, not from a point, but from a very small area δs (Fig. 14.3). At the point B_1 the solid angle ($\delta\Omega$) through which radiation

[1] See Appendix II (p. 230).

14. PHOTOSYNTHESIS, THERMODYNAMIC EFFICIENCY AND ATP

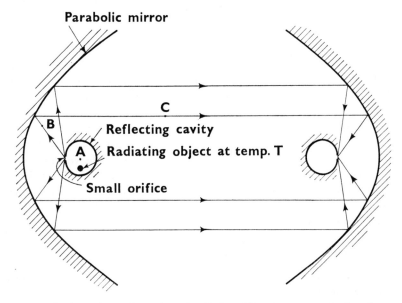

FIG. 14.2. Isothermal transformations of radiation. All surfaces are perfectly reflecting.

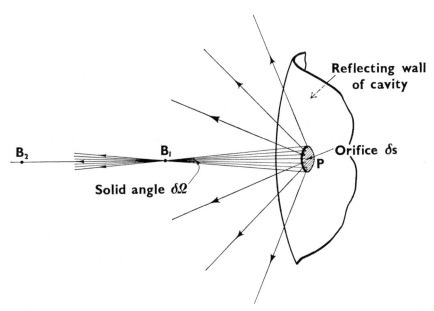

FIG. 14.3. Isothermal expansion of radiation.

is passing is the angle subtended by δs at B_1, since rays come from all over this area to pass through this point. At the point B_2, the luminous intensity has fallen off according to the inverse square law; but so too has the solid angle, $\delta\Omega$ in which the radiation arrives. Thus the intensity per unit solid angle is constant in the divergent beam. Free expansion of radiation (without absorption of course) from what can be regarded as a point source (such as the distant sun) is evidently accompanied by no change in its temperature, and the same obviously applies to reflection from a perfectly reflecting plane mirror. However, reflection from an ideal *white* surface (that is, one which absorbs no energy but merely reflects it diffusely), inasmuch as it at once increases the solid angle of the rays though it leaves their energy content unaltered, may lower its temperature very appreciably. This is of course one way of discussing why one can view a pure white surface in direct sunlight without discomfort, while to look steadily at the reflection of the sun in a mirror is impossible.

The Temperature of Monochromatic Radiation

Suppose at a point P (Fig. 14.4) in air or vacuum, monochromatic radiation is arriving uniformly through a small[1] solid angle Ω.

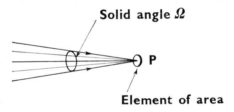

Fig. 14.4.

If I is the amount of radiant energy passing per second per unit area at P, the area being drawn normally to the axis of the cone, then the temperature of the radiation is given by[2]

$$T = \frac{h\nu}{k} \frac{1}{\ln\left(\dfrac{2h\nu^3 \Omega}{c^2 \mathscr{I}} + 1\right)} \qquad (14.2)$$

where h, k and c have their usual meanings (see table of constants on p. xii); ν is the frequency of the radiation, $\delta\nu$ being its bandwidth; and $\mathscr{I} = \dfrac{I}{\delta\nu}$ is the

---

[1] If the solid angle is large the intensity I must be defined as the amount of radiant energy falling on the surface of a sphere, centre P, whose great circle is of unit area.

[2] Planck, M. (1959). 'The Theory of Heat Radiation', p. 177. Dover Publications, New York; Constable, London.

14. PHOTOSYNTHESIS, THERMODYNAMIC EFFICIENCY AND ATP

specific intensity of the radiation in energy units per unit time per unit area per unit frequency interval.

In using this formula the following points should be noted. The constants h, k and c will have units as given on p. xii. The frequency will be in seconds^{-1} and for the red line at 6,800 Å will have the value (remembering that $c = \nu\lambda$)

$$\nu_{6800} = \frac{2 \cdot 998 \times 10^{10}}{6{,}800 \times 10^{-8}} = 4 \cdot 4 \times 10^{14} \text{ seconds}^{-1}.$$

The specific intensity, $\mathscr{I}$ will be the energy flux in ergs per second per centimetre^2 divided by the frequency width (in seconds^{-1}) of the band for which the measurement is made; its units will thus be ergs per centimetre^2. The term unity in the denominator of (14.2) is for visible light usually very small in comparison with the term $\left(\dfrac{2h\nu^3\Omega}{c^2\mathscr{I}}\right)$, and can consequently be neglected.

The formula as written refers to a medium of refractive index unity; but when a pencil of light enters water or some other denser medium its temperature (apart from any absorption or reflection) remains the same, so that the formula, if applied to the beam in the air just outside the medium, is still valid for the radiation within the medium.

Change of Temperature on Scattering

As an example of the use of this formula, suppose the change in the temperature of solar radiation on isotropic scattering (that is, scattering into a solid angle of 4π) is to be calculated. The sun has a diameter of about 864,000 miles and its distance from the earth is about 93,000,000 miles; consequently it subtends at the earth a solid angle of about

$$\frac{\frac{\pi}{4} \times 864{,}000^2}{93{,}000{,}000^2} = 0 \cdot 68 \times 10^{-4} \text{ steradian}[1]$$

and this is to be increased to 4π. Let the initial temperature T_1 of the radiation be 6,000°K. Then rearranging equation (14.2) somewhat we may write

$$\ln\left(\frac{2h\nu^3\Omega_1}{c^2\mathscr{I}}\right) = \frac{h\nu}{kT_1}$$

$$\ln\left(\frac{2h\nu^3\Omega_2}{c^2\mathscr{I}}\right) = \frac{h\nu}{kT_2}.$$

Subtracting these equations gives

$$\ln\left(\frac{\Omega_2}{\Omega_1}\right) = \frac{h\nu}{k}\left(\frac{1}{T_2} - \frac{1}{T_1}\right)$$

[1] See Appendix I (p. 230).

from which, writing $\Omega_1 = 0{\cdot}68 \times 10^{-4}$, $\Omega_2 = 4\pi$, $\nu = 4{\cdot}4 \times 10^{14}$ and $T_1 = 6{,}000°\text{K}$, we find
$$T_2 = 1{,}350°\text{K}$$
or about 1,080°C.

It should be noted that if the original radiation enters a nearly closed vessel with not totally absorbing walls repeated internal reflections will build up its temperature towards the initial value (Fig. 14.5). This is because

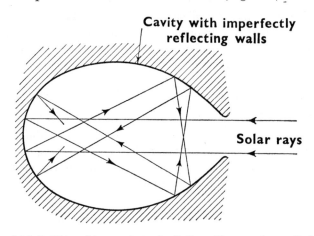

Fig. 14.5. Build up of temperature of radiation with successive scatterings.

such repeated reflections and scatterings raise the value of $\mathscr{I}$; but the process has a limit short of T_1 since eventually radiation will escape through the opening as fast as it enters. In practice, poor reflectivity and absorption will usually limit the temperature to far short of T_1.

Entropy of Radiation

Since the property of having both energy and temperature can be attributed to radiation it would seem only natural that it ought to have entropy, and this is indeed the case. In order to find out what its entropy is, use must be made of an imaginary mechanical device even more divorced from the ordinary run of common experience than the perfect gas engine; but in postulating such a device we have adduced reasons for believing that we are not contravening limits set by the fundamental constitution of nature, any more than when imagining frictionless bearings to be used. The use, in the present context, of perfectly reflecting surfaces is to be understood in this light.

Imagine a frictionless piston and cylinder enclosing a vacuous cavity. All the internal walls are perfect reflectors. Though this usually means that they are specular or mirror-like, in the present case they may also be white, the difference being that white surfaces give a *diffuse* reflection (Fig. 14.6); in neither case does the radiation suffer any loss of energy. Within the vacuum there is a non-volatile material object at temperature T (Fig. 14.7), the volume of the empty space around it being V. When the whole system is in equilibrium the space will be filled with isotropic[1] radiation to which must be attributed the temperature T. If the cavity was empty of radiation when the object was placed in it only those wavelengths will be present which the

Fig. 14.6. Reflection of radiation (a) specular, and (b) diffuse.

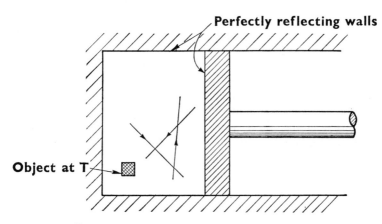

Fig. 14.7. Determination of the entropy of radiation.

object is capable, in any degree, of absorbing (and so emitting). Ordinarily this will mean that the cavity contains black-body radiation, even if the body is merely 'greyish'; but if a body is postulated which is perfectly transparent or reflecting to certain wavelengths then, by Kirchhoff's Law, these wavelengths will be absent. Thus monochromatic radiation can be conceived of as being present at equilibrium if necessary. What follows, however, is equally applicable to any mixture of wavelengths having the same temperature; it presupposes only equilibrium between the solid matter and the radiation. Incidentally, where the radiation is not complete black-body radiation

[1] It will also be unpolarized, but the question of polarization has been left out of our discussion as involving a slight, but undesirable, extra degree of complication.

the equilibrium is metastable, and the introduction of a minute speck of black material would effect a very rapid irreversible transformation, much as a small crystal would 'seed' a supersaturated solution.

Next imagine the piston to be very slowly withdrawn to increase the volume by dV. It is not difficult to appreciate that in order to maintain radiative equilibrium what must be kept constant is the density of the radiation in terms of energy per unit volume. If the total energy of the radiation is U, then the energy density u will be given by

$$u = \frac{U}{V} \qquad (14.3)$$

and an additional amount of energy dU equal to $u\,dV$ must appear in the radiation[1]. This comes from the solid body; and since this does no work it must, quite understandably, give up energy as heat. The whole process has been conducted reversibly; consequently the body has suffered a loss of entropy equal to dU/T. Unless the Second Law is inapplicable here, and the entropy of the universe can decrease, the radiation *must be regarded as acquiring an increase of entropy equal to this*, so that the net change of entropy (the process having been conducted reversibly) may be zero. If one thinks of radiation as composed of particles called photons not *altogether* unlike the molecules of a gas it is not difficult to appreciate that it can possess entropy as well as energy, though the analogy cannot be taken very far.

As a matter of fact the above expression for the entropy changes is not quite the whole story. Both theory and experiment reveal that radiation exerts a pressure, equal in fact to $\tfrac{1}{3}u$ when the radiation is isotropic. Consequently when the piston moves not only is heat energy required to create new radiation, but a further amount is needed to balance, as required by the First Law, the work done on the piston. This work is equal to $\tfrac{1}{3}u\,dV = \dfrac{dU}{3}$. Thus the heat lost by the object is altogether $\tfrac{4}{3}dU$, and the correct expression for the entropy of the radiation becomes

$$S_r = \frac{4}{3}\frac{U}{T} \qquad (14.4)$$

where the differential form can be dropped as the increment of radiation is of exactly the same quality as all the rest. This expression holds for radiation which is isotropic and unpolarized; but it may be of any wavelength or mixture of wavelengths, provided the appropriate temperature is used for each.

Spontaneous Processes in Radiation

By way of making matters clearer consider briefly, in the light of the above discussion, some spontaneous processes. When suitably imprisoned

[1] In other words, additional radiation must be formed to fill the volume dV to the required density.

14. PHOTOSYNTHESIS, THERMODYNAMIC EFFICIENCY AND ATP

monochromatic radiation is converted to black-body radiation by introducing a speck of soot the total energy of the radiation remains the same; but its temperature falls, since the energy in the original wavelength has to be shared amongst many. The entropy therefore rises (by equation 14.4), and this corresponds to the spontaneity of the process.

Again, when a parallel beam is reflected specularly by a perfect mirror there is no change in either the energy or the entropy; but when it is scattered by a perfectly white surface, although the energy is unaltered the temperature falls sharply, and this again implies an increase in the entropy. The principle underlying both these examples is the same: the introduction of an extra degree of disorder (whether by multiplication of wavelengths or of spatial directions) means an increase in the entropy.

'Economic' Efficiency in Photosynthesis

It is now possible to calculate what has been called the maximum economic efficiency with which the energy of radiation can be used in photosynthesis. We confine attention to a single simple case: isotropic radiation of energy, U_r and temperature, T_r falls on a leaf of temperature, T_1. The problem is, what fraction of U_1 can be equated to the free energy fixation in the leaf?

Suppose that not all of U_r is realizable as work. The balance may then be regarded as being rejected as heat (Q) by the leaf to its surroundings. Further, since the leaf obviously cannot be quite surrounded by an environment at its own temperature[1] it will lose unbalanced radiation (U_1) of its own while it is absorbing U_r. Thus if W is the work done at the expense of the radiation, then by the First Law

$$W = U_r - U_1 - Q. \tag{14.5}$$

Now for maximum efficiency there must be no overall entropy increase, so a balance sheet can be drawn up as follows:

Item	Entropy change
Radiation destroyed (U_r)	$-\dfrac{4}{3}\dfrac{U_r}{T_r}$
Radiation created (U_1)	$+\dfrac{4}{3}\dfrac{U_1}{T_1}$
Heat acquired by surroundings (Q)	$+\dfrac{Q}{T_1}$

Thus the limiting relation is:

$$-\frac{4}{3}\frac{U_r}{T_r} + \frac{4}{3}\frac{U_1}{T_1} + \frac{Q}{T_1} = 0. \tag{14.6}$$

[1] Naturally, it cannot be on the side facing the source of radiation.

Equations (14.5) and (14.6) give all that is needed, but before they are used matters can be simplified by noting that if T_r is appreciably greater than T_1 (and this will normally be true) then U_1 and U_1/T_1 will be negligible compared with U_r and U_r/T_r. This follows from Stefan's Law (see Appendix III). Neglecting these first two quantities gives therefore

$$W = U_r - Q \tag{14.7}$$

and

$$\frac{Q}{T_1} = \frac{4}{3}\frac{U_r}{T_r}. \tag{14.8}$$

Eliminating Q between these equations gives

$$W = U_r\left(1 - \frac{4}{3}\frac{T_1}{T_r}\right)$$

or

$$\eta_{\text{max.}} = \frac{W}{U_r} = 1 - \frac{4}{3}\frac{T_1}{T_r} \qquad (T_r \gg T_1) \tag{14.9}$$

which is the required result.

The Leaf as a Heat Engine

This result suggests that the leaf functions substantially as a heat engine; radiation is, after all, one of the modes of *heat* transfer. It was seen earlier (p. 215) that the maximum efficiency of a perfect heat engine is given by

$$\eta_{\text{max.}} = \left(1 - \frac{T_2}{T_1}\right) \tag{14.10}$$

which is very close to equation (14.9). The slight difference introduced by the factor 4/3 is due to the fact that the leaf does not merely absorb high-temperature heat; it also destroys radiation, but this is a small point. So also is the fact that the leaf delivers its work not in the classical form associated with mechanical forces, but as chemical free energy. Since it functions continuously the leaf-engine must employ a working substance which goes repeatedly through cycles, and certain of the electrons of the chlorophyll molecule may well fill this role. Naturally the size of the working elements of the engine means that quantum phenomena will be important.

The Quantum Efficiency of Photosynthesis

Formula (14.9) is relevant to the much-disputed question as to how many light quanta are essential, as a minimum, for the fixation of one molecule of carbon dioxide. Working in *moles* the question will be how many *einsteins* of light are required. If it is assumed that red light of wavelength 6,800 Å or frequency 4.4×10^{14} sec^{-1} is involved, the einstein has the value

$$E = h\nu \times \text{Avogadro's number}$$
$$= (6.625 \times 10^{-27}) \times (4.4 \times 10^{14}) \times (6.023 \times 10^{23})$$
$$= 1.75 \times 10^{12} \text{ ergs per mole}$$
$$= 42 \text{ kilocalories per mole approximately.}$$

If this is supplied as solar radiation scattered isotropically of resultant temperature 1,350°K (as in the example worked out previously), and if the temperature of the leaf is 27°C (or 300°K) the maximum work which can be derived from an einstein is equal to

$$42 \times \left(1 - \frac{4}{3}\frac{300}{1350}\right) = 29.5 \text{ kcal.}$$

Further, since one mole of glucose (equivalent to six of carbon dioxide) differs from its inorganic components by typically about 708 kcal of free energy (Chapter 11), the lowest quantum number for photosynthesis thermodynamically possible is in these circumstances

$$\frac{708}{6 \times 29.5} = 4.0 \text{ per molecule of } CO_2.$$

If the light intensity (and so T_r) is lower (and the desirability of this has often been stressed in experiments to determine the quantum number) the efficiency of conversion of the radiant energy will be *reduced*, and the minimum quantum number correspondingly increased.

Thermodynamics of High-Energy Phosphates

Since photosynthesis is directly concerned in the provision of high energy phosphate compounds this would seem an appropriate place to discuss their thermodynamic aspects. There are a large number of organic phosphates which play a part in the metabolism of living cells, and they might be described by the general formula R—OPO(OH)$_2$ where —OPO(OH)$_2$ is the phosphate radical. When such an organic phosphate is hydrolysed to form orthophosphoric acid:

$$R\text{—}OPO(OH)_2 + H_2O \rightarrow ROH + H_3PO_4$$

there is usually quite a large decrease in free energy, i.e. the reaction would go very spontaneously under the influence of the appropriate phosphatase. Such reactions naturally all have their own standard free energy changes, or standard affinities; and the value is found to depend quite markedly on the nature of R. For the so-called energy-rich compounds the bond between R and the phosphate radical is in fact a weak one; not much energy (or strictly, free energy) is needed to break it, and this leaves a larger excess of free energy when the two fragments (R— and —OPO(OH)$_2$ are joined to the elements of water (—OH and H—). Where the original bond is stronger it requires more free energy to break it and the excess left over is smaller. A high energy compound is thus an unstable one, like an explosive. As in all such cases, it will be unstable, either because there is a large energy decrease on hydrolysis ($-\Delta H$), or because there is a large entropy increase (ΔS). As a matter of fact as between one organic phosphate and another there are at least three elements in the overall hydrolysis which will allow differences; the splitting of the R—phosphate bond, the formation of the R—OH one, and changes in the ionisation constants of the acidic hydrogens of the phosphate groups.[1] The location of the free energy change actually in the original phosphate bond is thus a rather special convention that needs to be kept in mind.

Incidentally, as an instructive aside, the reason why in the previous paragraph the *free energy* required to break a bond had strictly to be referred to and not simply the *energy*, is that an entropy increase can mediate the breakage as well as an energy decrease. Of course, energy in the narrow sense is always required, but where there is an entropy increase, heat energy from the surroundings can be used to do the work, as was seen in Chapter 7; and there is always a superabundant supply of *that* available.

The Natural Philosophy of ATP

The convenience of having a commodity called 'currency' is so great that it is met with in nearly all communities, whether civilized or not. It enables purchasing power to be organized very flexibly and rapidly; and it is hardly surprising in view of this that it has been pressed into service as a metaphor, and that adenosine triphosphate has been called the 'universal energy currency' of living things. One feature of currency is that assets in the form of money can be used at once for a wide variety of purposes and in a wide variety of situations, such as paying for dinner or buying a railway ticket. If assets were in the form of farmland or furniture, matters could not be dealt with so expeditiously. It has often been remarked that ATP is, in fact, the currency used to pay, energetically speaking, for many widely different services; the synthesis of polymers, the contraction of muscle, and the doing of osmotic work, to name only a few. Another very striking implication of the use of ATP has however been pointed out[2]; it can be used equally well to

---

[1] Change in the acid strength of the phosphate groups is apparently an important and often overlooked factor (see George, P. and Rutman, R. J. Proc. Fifth Internat. Cong. Biochem. Vol. V, Pergamon Press, 1963).

[2] For full reference see footnote one, p. 166.

transfer energy in situations which differ widely in their redox potentials. Since in the vast majority of living things metabolic energy is ultimately obtained from oxidation, the oxidation-reduction urge or potential is a very important property, and it may vary considerably from point to point, as from the outside of a massive organ to the inside, or from arterial blood to venous blood. This means that if free energy were stored in a substance capable of being oxidized, transfer of the substance to another point of differing redox potential might leave it with much less free energy to give, since oxidation at the new site might not be such an exergonic process[1]. This disability however does not apply to ATP, since its energy is associated not with oxidation-reduction processes but with hydrolytic ones. The insensitivity to situation really follows from the fact that while the chemical potential of oxygen varies greatly from point to point in the organism that of water does not. This is suggested by the Gibbs-Duhem equation (compare equation 13.10) which for changes in an open system of many components at constant temperature and pressure can be written

$$x_1 \, d\mu_1 + x_2 \, d\mu_2 + x_3 \, d\mu_3 + \ldots = 0 \qquad (14.11)$$

where $x_1, x_2, \ldots$ are the mole fractions of the chemical species present, and $d\mu_1, d\mu_2, \ldots$ are the corresponding changes in their chemical potentials consequent on a small change in composition. The Gibbs-Duhem equation thus suggests that the chemical potential of a substance (like water) present in large amounts is much less sensitive to changes in composition than the chemical potential of a substance (like oxygen) present in small amounts[2]. That is fundamentally why ATP retains its energy value more consistently from point to point than would an oxidizable substance. In fact, in view of the consideration just outlined it seems entirely logical that living things should use an energy currency based on a hydrolytic reaction. Perhaps it is not too wild a guess to attribute to the phosphate radical, with its high stability, the role of a metal sufficiently untarnishable to form the exchangeable coin.

Some further matters probably unfamiliar to biologists but of interest in connection with the preceding chapter are discussed briefly below.

Appendix I: Measurement of Solid Angles

The concept of a solid angle is very similar to that of an angle in two dimensions. The latter is measured (Fig. 14.8) by the ratio of the arc (l)

---

[1] That is, a process for which ΔG is negative, and free energy is released.
[2] To make this clearer consider the case of only two components, say water and oxygen.

intercepted between the arms of the angle to the radius (r) for any circle drawn with centre at the vertex; that is,

$$\theta = l/r. \tag{14.12}$$

Similarly, the measure of a solid angle is the ratio of the surface (s) cut out by the cone forming the angle to the square (r^2) of the radius for any sphere whose centre is at the vertex of the cone; that is

$$\Omega = s/r^2. \tag{14.13}$$

Thus a complete circle has an angular measure of 2π *radians* and a complete sphere one of 4π *steradians*.

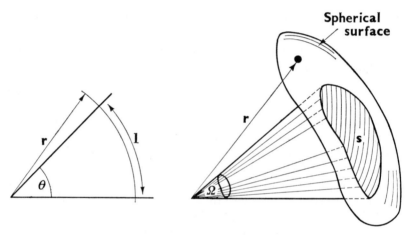

Fig. 14.8. Measurement of ordinary and solid angles.

Appendix II: Planck's Radiation Formula

The smallest indivisible element of radiation, the photon, has an energy (ϵ) called the quantum which depends on its frequency (ν) according to Planck's relation

$$\epsilon = h\nu \tag{14.14}$$

or

$$\epsilon = \frac{hc}{\lambda}$$

where c is the velocity of light and λ is the wavelength. Naturally, velocity and wavelength must both refer to measurements in the same medium, normally a vacuum or air. (The symbol c refers conventionally to the velocity *in vacuo*.)

Planck's formula for black body radiation, that is radiation whose spectral distribution and energy density per unit volume of space are such that it is in thermal equilibrium with matter, is as follows:

$$u_\nu = \frac{8\pi h \nu^3 n^3}{c^3} \left(\frac{1}{e^{h\nu/kT} - 1} \right) \quad (14.15)$$

where n is the refractive index of the medium (normally unity) and u_ν is the energy of frequency ν per unit frequency interval per unit volume. (Thus in a volume V the total energy in a band of width $\delta\nu$ would be $Vu_\nu\delta\nu$.)

From this formula the equation giving the temperature of beamed monochromatic radiation can be derived as equation (14.2).

Appendix III: Stefan's Law

This can be derived from Planck's Law. It expresses the relation between the total energy density of black-body radiation of all wavelengths and the temperature; or alternatively the rate at which a black body emits radiant energy. The relation in simple terms is

rate of energy radiation $\propto T^4$.

Thus the entropy of black-body radiation can easily be seen to vary as T^3, and the entropy radiated by the leaf will therefore normally be negligible compared with that absorbed with the incident radiation.

CHAPTER 15

The Thermodynamics of Irreversible Processes[1]

'*Is* there any more?' asked Pooh quickly.
Rabbit took the covers off the dishes, and said, 'No, there wasn't.'
'I thought not,' said Pooh, nodding to himself. 'Well, good bye. I must be going on.'
  So he started to climb out of the hole. He pulled with his front paws, and pushed with his back paws, and in a little while his nose was out in the open again . . . and then his ears . . . and then his front paws . . . and then his shoulders . . . and then—
'Oh, help!' said Pooh. 'I'd better go back.'
'Oh, bother!' said Pooh. 'I shall have to go on.'
'I can't do either!' said Pooh. 'Oh, help *and* bother!'

WINNIE-THE-POOH
A. A. Milne

This final chapter will be devoted to a development of thermodynamics which, though of fairly recent origin (the Norwegian physicist, Onsager published his famous paper in 1931), has already made many important contributions to biology. The Thermodynamics of Irreversible Processes, as it is often called, builds on the foundation of classical thermodynamics, but it uses the methods of statistical thermodynamics to establish its fundamental theorem. This theorem is derived from a principle of virtually universal generality, and as it therefore transcends all questions of mechanism it takes its place worthily alongside the other great principles of thermodynamics. Before discussing what it is, however, some preliminary points must be made clear.

SCOPE OF IRREVERSIBLE THERMODYNAMICS

Dealing as it does with the classical concepts of temperature, pressure and entropy (which are only precisely defined at equilibrium), it may puzzle the reader to see how irreversible thermodynamics can handle systems which are by presupposition out of equilibrium. It is one thing to apply, tentatively, exact thermodynamic equations to systems *accidentally* out of equilibrium and hope to find approximate agreement; but it seems quite another to start off with systems *intentionally* out of equilibrium and expect thermodynamics to lead to exact relationships. To put this in another way, since the new theorem to be introduced presupposes non-equilibrium, how can it legitimately be combined with the relationships of classical thermodynamics which

---

[1] General references:
  Prigogine, I. (1962). 'Introduction to the Thermodynamics of Irreversible Processes.' Wiley, New York.
  De Groot S. R. (1952). 'Thermodynamics of Irreversible Processes.' North Holland Publishing Co., Amsterdam.

presuppose equilibrium? On the face of it, it certainly does not look as if anything theoretically precise could result. However, this paradoxical situation does not, as might be supposed, imply an essential incompatibility at the heart of things. It is resolved by the understanding that the formulae of irreversible thermodynamics are theoretically exact only when the system is *indefinitely close to equilibrium*, at which position it can be said that the system is both in equilibrium and out of it. Having thus settled what might have remained with the reader as an unrecognized source of perplexity, it can be said at once that these formulae have a useful range of practical validity for quite appreciable deviations from equilibrium, though how large these deviations may be is a matter for actual experience to determine in each case. In general they are larger for transport processes than for chemical ones.

The theory leads to equations which describe steady states; that is, states which like true equilibrium are independent of time, but in which macroscopic processes are occurring. For this reason the subject is sometimes called the Thermodynamics of the Steady State. The role played by the concept of the steady state will be best understood however when we have discussed an example.

The Phenomenological Equations

First of all a method has to be devised of describing, in a simple way, how a system is changing. This is done by means of what are called phenomenological[1] equations; equations which limit themselves to observable quantities and do not attempt to plunge beneath the surface and deal with things in terms of such fundamental entities as atoms, molecules and electrons. The reader will already be familiar with many simple phenomenological equations; Ohm's Law and Fick's Law of diffusion are familiar examples. The phenomenological equations of irreversible thermodynamics are very like these, but more general. The necessity for being more general is not difficult to appreciate. Fick's Law of diffusion

$$\frac{\mathrm{d}m}{\mathrm{d}t} = \mathscr{D}\frac{\mathrm{d}c}{\mathrm{d}x} \tag{15.1}$$

(where $\frac{\mathrm{d}m}{\mathrm{d}t}$ is the rate of passage of solute across unit area under the concentration gradient $\frac{\mathrm{d}c}{\mathrm{d}x}$ and $\mathscr{D}$ is the diffusion coefficient) may be adequate in a simple system; but suppose that while one solute is diffusing under its concentration gradient another solute is simultaneously moving under *its* gradient: can it be assumed that the movement of the second is quite without influence on that of the first? Or suppose that while a solute is diffusing there

[1] From the Greek word 'to appear'.

is simultaneously a temperature gradient in the system with a consequent flow of heat: is the flow of heat a matter of indifference to the solute, so that the movement of the latter is still adequately described by (15.1)? The answer to these questions, from experience, theory, and one might almost add intuition, is no. Equation (15.1) often ceases to be adequate, and something more general is needed.

Among various possible approaches it is found convenient and adequate to generalize equation (15.1) (confining ourselves to the simple one-dimensional case) to

$$\frac{dm_1}{dt} = \mathscr{D}\frac{dc_1}{dx} + \mathscr{E}\frac{dc_2}{dx} \qquad (15.2)$$

which allows the incorporation by a new coefficient, $\mathscr{E}$ of a possible effect of the gradient of a second solute on the rate of movement of the first. Of course there will be another equation similar to (15.2) for this second solute:

$$\frac{dm_2}{dt} = \mathscr{F}\frac{dc_2}{dx} + \mathscr{G}\frac{dc_1}{dx}. \qquad (15.3)$$

Should there be still further solutes, or should other entities like heat be moving, extra terms will need to be added to take account of them in a similar way.

Onsager's Reciprocal Relations

Before proceeding, there is scope for simplifying and rationalizing the notation. In the first place a single symbol (say J) can be used for $\dfrac{dm}{dt}$, and $J_1, J_2 \ldots$ can be written for the rates of movement of the different entities (solutes, heat, etc.) concerned. Then, to assist the memory further the multiplicity of coefficients $\mathscr{D}, \mathscr{E}, \mathscr{F} \ldots$ can be replaced with a single symbol L with subscripts. With this notation $\mathscr{D}$ becomes L_{11}, and $\mathscr{E}$ becomes L_{12}, the rule being that the first subscript refers to the component moving (and is thus the same as the subscript on the corresponding J); the second refers to the component whose gradient is being taken into account. Thus the equation describing the movement or 'flux' of component (1) becomes

$$J_1 = L_{11}X_1 + L_{12}X_2 + L_{13}X_3 + \ldots \qquad (15.4)$$

where in addition, X has been used with the appropriate subscript to refer to the gradient or 'force'[1]; that is $X_1 = \dfrac{dc_1}{dx}$, $X_2 = \dfrac{dc_2}{dx}$, and so on. This notation is both convenient and easy to remember. In general terms J_i is

---

[1] The term 'force' is clearly used here in a much wider sense than in mechanics. This wider use has become quite general, though 'affinity' is also sometimes used.

the flux of i, X_i is the corresponding force or gradient promoting movement of i, and L_{ij} is the coefficient describing how the flux of i is influenced by the gradient of j.

The fundamental theorem of irreversible thermodynamics can now be stated. Onsager showed in 1931 by means of statistical theory that provided the fluxes and forces are measured in a particular way, two L-coefficients with the same suffixes are equal. In symbols,

$$L_{12} = L_{21}$$

or $$L_{ij} = L_{ji}. \qquad (15.5)$$

Expressed in words this means that the coefficient which expresses the influence of the force j on the flux of i is the same as the coefficient which expresses the effect of the force i on the flux of j. The particular way in which fluxes and forces are measured will have to be discussed in a moment, but firstly two comments are necessary. Ohm's Law is usually written

$$E = IR$$

or $$I = \frac{1}{R}.E \qquad (15.6)$$

where E is the potential difference, I the current, R the resistance, and $\frac{1}{R}$ the conductance. Comparison of equation (15.6) with equation (15.4) shows that the L's can be regarded as 'conductance' coefficients. Like electrical conductance, their magnitude will depend on such things as the dimensions and temperature of the system (on the macroscopic level) and on the activation energies involved in the processes (on the molecular level). Thus doubling the physical size of the system will double the L's; and in biological systems the presence of enzymes will have a profound effect on them.

Secondly, Onsager's reciprocal relation (15.5) is a consequence of the fact that it makes no difference to the fundamental equations of motion (quantum as well as classical Newtonian) if t is replaced by $-t$. Time has therefore no direction so far as these equations are concerned. If a cinematograph film could be taken of molecular events in an 'aged' system (one which has been long at equilibrium), then it might be projected backwards instead of forwards and no one would be any the wiser; it would be equally valid as a record of physically probable sequences. This is not of course true with regard to systems not in equilibrium, where the Second Law distinguishes between the two directions of time in a most definite manner. In practice, while Onsager's relations were originally proved only for deviations from equilibrium small enough to be treated as spontaneous statistical fluctuations, they are found to lead to results verified experimentally and in a few cases by kinetic theory, when applied to departures from equilibrium very much more than these.

Measurement of Fluxes and Forces

It goes without saying that if a quite general result like equation (15.5) is to be obtained, the fluxes and forces cannot be measured just anyhow; a quite general method of specifying them is required. It will hardly surprise the reader that such a method is found in connection with the concept of entropy, since this concept supplies the only measure we have of all the various

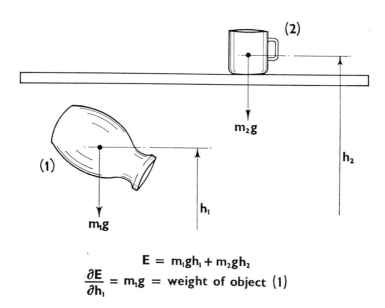

$$E = m_1 g h_1 + m_2 g h_2$$
$$\frac{\partial E}{\partial h_1} = m_1 g = \text{weight of object (1)}$$

Fig. 15.1. A mechanical analogy.

types of irreversible change associated with the movement towards equilibrium. In order to see how this suggestion is developed consider a simple gravitational analogy, a number of objects on an elevated shelf (Fig. 15.1). The potential energy E of this system can be written

$$E = m_1 g h_1 + m_2 g h_2 + \ldots \tag{15.7}$$

If the shelf is shaken so that the objects fall off the system proceeds to move to a new position of equilibrium (with the objects probably on the floor) characterized by a lower value for the potential energy. If equation (15.7) is differentiated with respect to time an equation is obtained which describes this movement to equilibrium:

$$\frac{dE}{dt} = m_1 g \frac{dh_1}{dt} + m_2 g \frac{dh_2}{dt} + \ldots \tag{15.8}$$

About this equation the quite general statement can be made that $\frac{dE}{dt}$ must be negative[1], and this marks it out as being more than a trivial one for the system. Further, it shows how, having split up the whole system into parts (each falling object constituting one part), the contribution each part makes to $\frac{dE}{dt}$ can be represented as the product of a 'force' mg (actually the weight of the object) and a 'flux' $\frac{dh}{dt}$.

Turning from this simple gravitational analogy to the more general thermodynamic situation, $\frac{dE}{dt}$ naturally has to be replaced by $\frac{dS}{dt}$, since the criterion for spontaneous change no longer concerns E, but rather the entropy S. Next, fluxes analogous to $\frac{dh_1}{dt}$, $\frac{dh_2}{dt}$, ... have to be chosen which will describe how the aspect of the system changes with time. There is a wide range of choices possible here, and the actual ones made will be dictated by considerations of interest and convenience. Having chosen these fluxes our liberty ends, for the forces to be employed with them are then settled by the requirement that the product of a flux and its force must represent the contribution that part of the overall process makes to the 'rate of entropy production' $\frac{dS}{dt}$. In symbols we shall have

$$\frac{dS}{dt} = \sigma = J_1 X_1 + J_2 X_2 + J_3 X_3 + \ldots \tag{15.9}$$

where a new symbol (σ) has been introduced and the fluxes have been written in front of the forces instead of after them as in the analogous equation (15.8). How equation (15.9) is used to choose suitable fluxes and then to indicate their corresponding forces will become clearer when we have discussed a simple but important application.

Method of Calculating the Entropy Production

Before proceeding further there is a general point of view to be taken. We are concerned with a system which is out of equilibrium and in which irreversible processes are therefore taking place. What we require is to obtain a general expression for the rate at which entropy is being produced *by those irreversible processes*. There are broadly two ways of doing this. In the first, we imagine the system to be thermally insulated, when of course the

---

[1] Strictly the system must be non-conservative (unlike a pendulum) for this to be true.

whole increase in its entropy must arise from the irreversible processes within it and represent a definite creation of entropy. In the second, we allow it to interact freely with its surroundings, exchanging heat, work and in some cases matter with them; then we disentangle from its total entropy change that part which has been caused by the processes within it. The other part we discard. It does not necessarily represent an entropy creation at all, since it may be counterbalanced by entropy lost by the surroundings; but even if this is not so it will be linked with irreversible processes other than the ones we are concerned with. That is the reason why we neglect it.

Needless to say both of these procedures yield the same result. For reasons of space only the first will be illustrated. Once the expression for σ is obtained it can be dissected into fluxes and forces as required.

In general, the calculation of σ requires the following principles to be used:

(1) the conservation laws (of mass, energy and electric charge especially);

(2) the general thermodynamic equation for an open system (see equation (10.16)):

$$dU = TdS - PdV + \Sigma \mu_i dn_i. \qquad (15.10)$$

This latter equation is used in the form expressing dU rather than dG (that is, equation (10.15)), since in the first place, U is subject to the law of conservation (a great convenience), and in the second, G is not generally of much use where temperature is non-uniform, as it will often be supposed to be. Equation (15.10), as will be remembered from the rather lengthy discussion of it, applies to a system in equilibrium, the addition of matter implied by the last term ($\Sigma \mu_i dn_i$) being made so infinitely slowly that internal equilibrium is never jeopardized. This equation will be applied to sub-systems where the influx of matter has, in fact, a distinctly measurable speed, just as an influx of heat embraced by the term TdS has. However, the justification for this is that on the practical level, it leads to results borne out by experiment; and on the theoretical, that there is reason to believe that the equation will be true within the limits of experimental accuracy as long as the disequilibrium is not violent enough to make nonsense of the notion of temperature.

Thermo-osmosis

The phenomenon chosen for consideration is that of thermo-osmosis, a phenomenon that has a superficial resemblance to osmosis but which results from a difference of temperature. The simplest possible case will be considered, a system containing only pure water and divided into two by a permeable membrane (Fig. 15.2). The two sub-systems (α and β) are homogeneous within themselves, but both temperature and pressure vary as between the compartments. Consider the whole system to be isolated, so that when the entropy production is calculated this can be referred wholly to the irreversible process occurring at the membrane[1], a process which will involve both flow of matter and transport of heat. Isolation will mean that the volume of the entire system remains constant so that no work is exchanged between it and the surroundings, and also that its boundaries are composed of thermally-insulating material.

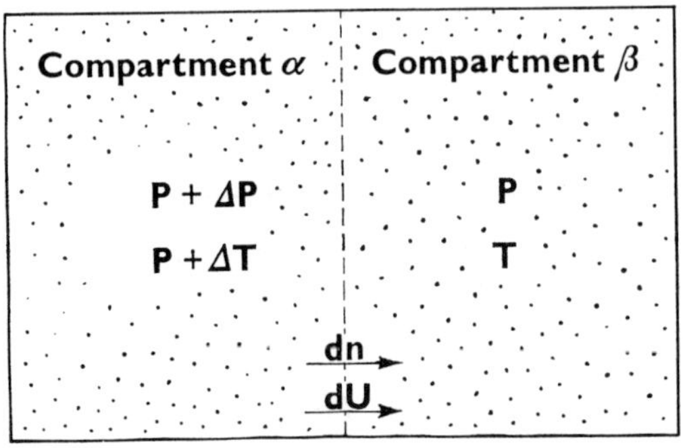

FIG. 15.2. A thermosmotic system. Both compartments contain water only.

We commence by considering what changes take place when a quantity of matter dn moles and a quantity of energy dU move from compartment α to compartment β. We choose matter and energy because they are conserved, and (under the condition of isolation) what one sub-system loses the other exactly gains. Had heat been chosen instead of energy this would no longer be true; work is involved when matter moves from a higher pressure to a lower, and this work will usually involve heat changes. Hence it is by no means a necessary consequence that the heat lost by one sub-section is equal to the heat gained by the other. As a matter of fact this puts the matter really rather too simply, for it is a difficult thing to say quite what one means, even in theory, by heat lost or gained when matter at differing temperature is also passing; but enough has been said to show the advantage of dealing with U rather than Q. Incidentally, the increments dn and dU are independent, each clearly being influenced differently by the pressure and temperature differentials. In the following analysis, superscripts α and β are used to denote the two sub-systems, and the symbol Δ to indicate the difference in some intensive variable between the latter; for example, $\Delta T = T^\alpha - T^\beta$.

[1] This would be true if the system is merely *thermally* insulated, since the exchange of *work* (slowly of course) does not change its entropy. But it is convenient to imagine the system to have constant volume too, as this makes its total energy constant and simplifies matters without altering the outcome.

15. THE THERMODYNAMICS OF IRREVERSIBLE PROCESSES

Denoting by dS^α the increment in the entropy of section α when the matter (dn) and energy (dU) leave it we may write

$$dS^\alpha = -\left[\frac{\partial S^\alpha}{\partial n}dn + \frac{\partial S^\alpha}{\partial U}dU\right] \tag{15.11}$$

which like equation (3.16) is a purely mathematical relation; at least the only thermodynamics it contains is the implication that S is a function of n and U. (The negative sign appears because dn and dU have not, for section α, their usual meaning of positive increments, but are both negative ones.) Similarly for section β

$$dS^\beta = \frac{\partial S^\beta}{\partial n}dn + \frac{\partial S^\beta}{\partial U}dU. \tag{15.12}$$

The increase in the total entropy (S) of the system is found by adding these two equations. The result can be written

$$dS = dS^\alpha + dS^\beta = dn\left(-\frac{\partial S^\alpha}{\partial n}+\frac{\partial S^\beta}{\partial n}\right) + dU\left(-\frac{\partial S^\alpha}{\partial U}+\frac{\partial S^\beta}{\partial U}\right)$$

$$= dn\varDelta\left(-\frac{\partial S'}{\partial n}\right) + dU\varDelta\left(-\frac{\partial S'}{\partial U}\right) \tag{15.13}$$

using the notation involving $\varDelta$, the dash $(')$ being used to denote that the symbol refers to a sub-system.

The values of $\dfrac{dS'}{dn}$ and $\dfrac{dS'}{dU}$ can be evaluated by noting that each sub-system is constant in volume, but can gain or lose matter. This enables equation (15.10) to be re-written in the form (putting $dV = 0$ and adding dashes to indicate that subsystems are being referred to),

$$dU' = TdS' + \mu dn. \tag{15.14}$$

In passing, it might be urged that constancy of volume was postulated only for the *entire* system, and therefore it might be allowed that the *separating membrane* deflects as water moves from α to β. However, if this were granted, remembering the difference of pressure which must be sustained, account would have to be taken of the energy stored elastically in the membrane. This is really no different from energy escaping into the surroundings[1]; hence if we wish to be able to regard the total energy U as fixed the membrane must be postulated as being rigid, i.e. we must allow no volume changes in the sub-sections either. Otherwise, a correction will have to be

---

[1] In fact, the containing vessel and the membrane are, thermodynamically, part of the surroundings. The irreversible changes being studied do not concern them.

made for the change in U.

Writing $\mathrm{d}U$ as zero in equation (15.14) gives at once

$$\frac{\partial S'}{\partial n} = -\frac{\mu}{T} \qquad (15.15)$$

and similarly putting $\mathrm{d}n$ zero gives

$$\frac{\partial S'}{\partial U} = \frac{1}{T}. \qquad (15.16)$$

Inserting equations (15.15) and (15.16) into the mathematical identity (15.13) gives the thermodynamic result

$$\mathrm{d}S = \mathrm{d}n \varDelta\left(\frac{\mu}{T}\right) + \mathrm{d}U \varDelta\left(-\frac{1}{T}\right) \qquad (15.17)$$

or dividing by $\mathrm{d}t$,

$$\sigma = \frac{\mathrm{d}S}{\mathrm{d}t} = \frac{\mathrm{d}n}{\mathrm{d}t}\varDelta\left(\frac{\mu}{T}\right) + \frac{\mathrm{d}U}{\mathrm{d}t}\varDelta\left(-\frac{1}{T}\right). \qquad (15.18)$$

This equation is the first stage towards the required result, and it is worth pausing for a moment to consider some of its features. It is obviously in the same form as equation (15.9), with the fluxes given by $J_m = \dfrac{\mathrm{d}n}{\mathrm{d}t}$ and $J_U = \dfrac{\mathrm{d}U}{\mathrm{d}t}$, and the forces X equal to the differences, as between sections, of the expressions $\left(\dfrac{\mu}{T}\right)$ and $\left(-\dfrac{1}{T}\right)$. It might have been expected that, as in the case of passive movement (Chapter 13), the force driving the water would somehow have involved the difference in chemical potentials (that is, $\varDelta\mu = \mu^\alpha - \mu^\beta$), and this is in fact almost realized. However, since an *entropy* increase is being calculated the dimensions would not have been right; the division by T corrects for this. Turning to the energy flow, intuition might have suggested that this would be proportional to $\varDelta T$ (though it must be recognized that more than simple heat flow is concerned). To have the dimensions right it would then be necessary to divide by T^2, and that is actually just how matters turn out, since[1]

$$\varDelta\left(-\frac{1}{T}\right) = \frac{\varDelta T}{T^2} \qquad (15.19)$$

$\varDelta T$ being small.

The last point to notice is that while equation (15.18) is perfectly straightforward, it is not a relationship that can be put to direct experimental test.

---
[1] Treating the symbol $\varDelta$ as a differentiation (see Appendix I, p. 265).

15. THE THERMODYNAMICS OF IRREVERSIBLE PROCESSES

This arises from the fact that U and μ, unlike n, are unknown to the extent of arbitrary additive constants. If the equation is ever to mean anything, physically, it must be possible to turn it into a form in which these arbitrary constants cancel out. This line of thought suggests that we try to modify the fluxes and forces by measuring them in rather different terms, so that the physically objectionable features disappear. As the fluxes are in a sense the primary quantities, and as $J_m = \dfrac{\mathrm{d}n}{\mathrm{d}t}$ is not an offender, this means that we must try to find a way of re-defining the second flux.

Change of Flux

The above suggestion can be developed in a way which seems intuitively sensible, and which leads to a system of fluxes and forces which are all unambiguously defined. When a quantity of matter, $\mathrm{d}n$ moles, passes from one section to the other it may be regarded as inevitably carrying with it a certain amount of energy. If U is the total energy of the system, and n moles its total content of matter, the energy per mole or molar energy can be written

$$u = \frac{U}{n} \qquad (15.20)$$

and the $\mathrm{d}n$ moles might be regarded as being bound up with an amount of energy $u\mathrm{d}n$. However, this is not the whole story. The $\mathrm{d}n$ moles has a volume $v\mathrm{d}n$, where v is the molar volume, and it seems clear that a further transfer of energy equal to $Pv\mathrm{d}n$ is inevitably involved. This statement can be justified by reflecting that one sub-system 'pushes out' a volume of water $v\mathrm{d}n$, and so does work $Pv\mathrm{d}n$ upon it; the other sub-system in the process of receiving the water has this work imparted to it. It might reasonably be asserted therefore that the $\mathrm{d}n$ moles has necessarily linked with it an amount of energy equal to

$$u\mathrm{d}n + Pv\mathrm{d}n = (u+Pv)\mathrm{d}n = h\mathrm{d}n \qquad (15.21)$$

where $h = \dfrac{H}{n}$ is the molar enthalpy of the water.

Now the quantity h is subject to an unknown additive constant. If $h\mathrm{d}n$ is therefore subtracted from the original $\mathrm{d}U$ this constant might possibly cancel out. Further, the quantity left over might legitimately be regarded as energy which passes on its own account, that is, not integrally associated with the flow of matter; it might in fact be regarded as a heat flow. This leads to the definition of a new flux by the relation

$$J_Q = J_U - hJ_m \qquad (15.22)$$

in which this suggestion is incorporated. From equation (15.22)

$$J_U = J_Q + hJ_m \tag{15.23}$$

and inserting this into equation (15.18) and rearranging we find

$$\sigma = J_m\left[\Delta\left(\frac{\mu}{T}\right) + h\Delta\left(-\frac{1}{T}\right)\right] + J_Q\Delta\left(-\frac{1}{T}\right) \tag{15.24}$$

with X_m changed but X_Q as before (that is, equal to X_U). Whether this equation is physically meaningful will depend on how the force X_m turns out.

To develop X_m equation (10.15)

$$dG = VdP - SdT + \Sigma\mu_i dn_i \tag{15.25}$$

is needed in a modified form. This equation, it will be remembered, is the general equation for an open system in equilibrium. The companion equation (10.16) in which dU appears instead of dG (see equations (15.10) and (15.14)) has already been used. We first regard the system to which it refers as closed[1], when the last term disappears; we then regard it as consisting of a pure substance, and divide it through by the molar content (n) of this substance. This gives a relation between differentials of the molar free energy (g), molar volume (v) and molar entropy (s), the small letters indicating that these *intensive* properties are now being dealt with. Finally, we note that for a pure substance the molar free energy $\left(g = \dfrac{G}{n}\right)$ is equal to the chemical potential (μ). This gives

$$dg = d\mu = vdP - sdT \tag{15.26}$$

or using the symbol Δ to indicate that the differential is actually between the two sub-systems (that is, a matter of a difference in space rather than in time),

$$\Delta\mu = v\Delta P - s\Delta T. \tag{15.27}$$

Further, in accordance with its definition we have

$$\mu = g = h - Ts$$

or

$$h - \mu = Ts. \tag{15.28}$$

X_m can now be simplified. Remembering that Δ is to be treated as a differential we have (see equation (15.24))

$$\Delta\left(\frac{\mu}{T}\right) = \frac{T\Delta\mu - \mu\Delta T}{T^2} \tag{15.29}$$

and

$$h\Delta\left(-\frac{1}{T}\right) = \frac{h\Delta T}{T^2}. \tag{15.30}$$

[1] This merely simplifies matters, as n can then be regarded as constant in the subsequent division.

Adding,
$$X_m = \frac{T\Delta\mu + (h-\mu)\Delta T}{T^2}$$

or introducing equations (15.27) and (15.28),
$$X_m = \frac{T(v\Delta P - s\Delta T) + Ts\Delta T}{T^2}$$
$$= \frac{v\Delta P}{T} \qquad (15.31)$$

which is a gratifying result, physically quite unambiguous and intuitively plausible. It may be remarked in this latter connection that vJ_m is the volume flow of matter; multiplied by ΔP this becomes a rate of doing work; and divided by T it assumes the correct dimensions.

The Phenomenological Equations

The fluxes can now be expressed in the phenomenological or 'Ohm's Law' form. They become

$$J_m = L_{mm}\frac{v\Delta P}{T} + L_{mQ}\frac{\Delta T}{T^2} \qquad (15.32)$$

$$J_Q = L_{Qm}\frac{v\Delta P}{T} + L_{QQ}\frac{\Delta T}{T^2} \qquad (15.33)$$

in which, since precautions have been taken to measure the fluxes and forces in the correct way we have the relation

$$L_{Qm} = L_{mQ}. \qquad (15.34)$$

The main physically interesting results come from applying these equations to particular simple cases. Although the theory was developed by assuming the system to be isolated, it would obviously make no difference to equations (15.32) and (15.33) if it were allowed to interact with the surroundings, say by receiving heat from them. The irreversible processes at the membrane only take cognizance of the momentary values of ΔP and ΔT, not of how these change or of how they are caused. Thus we may assume that experimental arrangements are such that each sub-system is maintained at the same temperature (that is, $\Delta T = 0$) by being in intimate contact with a suitable heat source or sink. Under these conditions equations (15.32) and (15.33) reduce to

$$J_m = L_{mm}\frac{v\Delta P}{T} \qquad (15.35)$$

$$J_Q = L_{Qm}\frac{v\Delta P}{T} \qquad (15.36)$$

or

$$\frac{J_Q}{J_m} = \frac{L_{Qm}}{L_{mm}} = Q^*. \tag{15.37}$$

This equation indicates that (provided the disequilibrium is small enough, a universal presupposition in the theory) the *isothermal* flow of unit quantity of matter is associated with the flow of $\dfrac{L_{Qm}}{L_{mm}}$ units of heat. In other words, it is to be expected that this quantity of heat has to be supplied experimentally to the first sub-system and removed from the second (per unit of water passing) to keep their temperatures constant. This quantity is called the 'heat of transfer' and is given the symbol Q^*. It is typical of a whole range of similar quantities in irreversible thermodynamics. It can be visualized very readily with reference to a simple system (Fig. 15.3) in which an osmotic potential

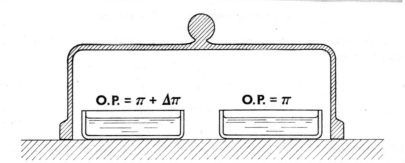

Fig. 15.3. A simple system where the heat of transfer Q^* is substantially the ordinary latent heat.

difference $\Delta \Pi$ replaces the pressure difference ΔP, this alteration introducing no significant change into the theory[1]. Under isothermal conditions, water will distil from the weaker to the stronger solution, and every unit which passes will convey its latent heat from one to the other. Provided the two solutions are nearly alike (as required by the theory) their latent heats will be virtually identical, and the heat supplied to the weaker solution to maintain its temperature will be exactly matched by that removed from the stronger; this makes the concept of a heat flow unequivocal. The heat of transfer, Q^* is in this case almost exactly equal to the latent heat, there being a small addition associated with the kinetic energy accompanying the 'drift' movement from one vessel to the other.

What the quantity Q^* implies is a 'characterization' of the membrane; up to the stage represented by equations (15.32) and (15.33) the nature of the

---

[1] The separating membrane is represented of course by the air space.

15. THE THERMODYNAMICS OF IRREVERSIBLE PROCESSES

membrane has been left entirely open. The value of Q^* for any particular example can in principle be determined experimentally, and in fact it is connected very simply with the temperature dependence of the 'permeability' as will be seen later (Appendix II, p. 265).

THE THERMO-OSMOTIC PRESSURE

To return to the discussion of equations (15.32) and (15.33), the second simple case of interest arises when the pressure and temperature differentials assume such values that the flow of matter is reduced to zero; this constitutes a 'steady state'. Writing equation (15.32) equal to zero and rearranging gives

$$\frac{\Delta P}{\Delta T} = -\frac{L_{mQ}}{L_{mm}}\frac{1}{vT} \tag{15.38}$$

or

$$\Delta P = -\frac{L_{mQ}}{L_{mm}}\frac{\Delta T}{vT}. \tag{15.39}$$

The benefit of the Onsager reciprocal relationship can now be reaped, for introducing equations (15.34) and (15.37) we find

$$\Delta P = -\frac{Q^*}{vT}\Delta T, \tag{15.40}$$

an equation expressing what might be called the thermo-osmotic pressure. It should be noticed in connection with this name, that in the steady state while J_m is zero, J_Q is not; consequently the thermo-osmotic pressure requires a continuous flow of heat, or dissipation of energy, for its maintenance[1]. However, this dissipation cannot be calculated without further information; in particular more must be known about the coefficients L_{mm}, L_{QQ} and L_{mQ}. The subject of this energy dissipation will be taken up later.

CASE OF LATENT HEAT

As the problem was mentioned in an earlier chapter (Chapter 13) we will pause to calculate, from equation (15.40) the case in which the membrane is an 'aerial' one and Q^* is very nearly the latent heat. We note first that equation (15.40) may be read with $-\Delta \Pi$ in place of ΔP, since equal values of both have the same influence on the change of chemical potential (cf. equation (13.11)). Writing $v = 0.001$ litre per gram[2], $T = 298°K$, $L = 582$ calories per gram $= 24.05$ litre-atmospheres per gram we have

$$\frac{\Delta P}{\Delta T} = -\frac{24.05}{0.001 \times 298}$$

$$= -80.6 \text{ atmospheres per degree,}$$

[1] In this it is in marked contrast with the *osmotic* pressure. The difference lies in the distinction between 'active' and 'passive'.
[2] Alternatively these quantities might have been expressed per *mole*.

the negative sign indicating that the higher pressure develops on the low temperature side. This confirms how influential a temperature gradient can be in offsetting a gradient of suction potential, particularly where, as in cells, the membrane may be lipoidal instead of aerial, and involve a higher value[1] of Q^*. Of course, a membrane may have quite a low value of Q^* and yet function as semi-permeable; there is no necessary simple proportionality between the osmotic and the thermo-osmotic effects.

The General Entropy Production

By using the flow equations and inserting them into equation (15.9) we can find an expression for the rate of entropy production in terms of the forces as the only variables. The result can be written

$$\sigma = L_{mm}X_m^2 + (L_{mQ}X_mX_Q + L_{Qm}X_QX_m) + L_{QQ}X_Q^2 \qquad (15.41)$$

the two middle terms being equal. By the Second Law, σ must always be positive, and this imposes a restriction on the coefficients, L. The algebraic expression $(ax^2 + 2hxy + by^2)$ can be written $a\left(x + \dfrac{hy}{a}\right)^2 + \left(b - \dfrac{h^2}{a}\right)y^2$. If this expression is to be essentially positive, the *coefficients* of the two square terms must both be positive, since the squares are necessarily so themselves. This means that a must be positive[2] (an intuitively obvious result), and the quantity $\left(b - \dfrac{h^2}{a}\right)$ must be positive. Thus we must have $ab > h^2$. In terms of the L's this means that experiment is bound to show that

$$L_{mm}L_{QQ} > L_{mQ}^2 \qquad (15.42)$$

but about the question of by how much it is greater, thermodynamics says nothing. Equation (15.42) contains the sole information given on this point.

The Criterion of Active Transport

Reference to equation (15.41) will show that of the three terms making up the expression for the entropy production the two end ones are necessarily positive. The forces appear as squares, and the two 'conjugate' coefficients L_{mm} and L_{QQ} (corresponding to the a and b in the algebraic expression above) are necessarily positive. The middle term may be either positive or negative, since the signs of the 'cross-coefficients', and of X_m and X_Q may have either sense. This provides just the criterion required for a definition, in the widest context, of what is meant by 'active transport' or an active process of a chemical

---

[1] That is, it may be 'harder' for water to 'evaporate' into a lipid phase than into an aerial one.
[2] And by symmetry also b.

15. THE THERMODYNAMICS OF IRREVERSIBLE PROCESSES

nature. When a complex overall process can be analysed into a number of independent movements or fluxes *those must be thought of as 'active' in which the cross-coefficients* (for example L_{mQ}) *are important*. This definition is comprehensive in that it embraces non-isothermal situations as well as the biologically more familiar isothermal ones, and allows that the active movement may be 'downhill' (that is, give a positive contribution to σ) as well as 'uphill' (the more spectacular and interesting situation for biologists).

Although the discussion has been limited to a simple system with two fluxes only, what has been said can be extended to a system of any complexity required. The Onsager relations (15.34) become more numerous, and so do such inequalities as (15.42). Chemical reactions can be taken account of as well as transport processes; but the main foundations of the theory remain the same.

The Energy Dissipation

Where irreversible processes are taking place, a capacity to do work is being irretrievably lost; in other words energy is being dissipated. When an unbalanced increase of entropy, dS, occurs the loss of useful work, in the context of the whole universe, can be shown to be $T_0 dS$, where T_0 is the lowest existent temperature. However, in the context of the local system where dS occurs T_0 must be replaced by the local temperature. Thus in connection with the theory of irreversible processes, the product $T\sigma$ can be spoken of as the rate of (local) dissipation of energy. It is instructive to look at the example we have studied in this light. The flux J_m represents a volume flow of matter equal to vJ_m. This is dropped in pressure from $(P+\Delta P)$ to P. Consequently, were it to be harnessed in a piston and cylinder it could be made to do work at the rate of $vJ_m \Delta P$. It should be noted that v is not assumed to change from one side of the membrane to the other, so that the work does not involve drawing on the internal energy of the matter in transport.

Concurrent with the above there is a flow of heat J_Q from a temperature $(T+\Delta T)$ to one of T. The maximum efficiency[1] with which this heat could be harnessed to give work is $\dfrac{\Delta T}{T}$, the corresponding rate of working being $J_Q \dfrac{\Delta T}{T}$. Thus the maximum rate at which the total irreversible process could be made to do work is given by the equation

$$\text{maximum rate of working} = (vJ_m \Delta P) + \left(J_Q \frac{\Delta T}{T}\right).$$

[1] See Chapter 14, where the corresponding expression is $(T_1-T_2)/T_1$, cf. equation (14.1).

However, all this potentiality is wasted since the system delivers nothing useful. Consequently this expression becomes the rate of energy dissipation, and comparison with equations (15.9), (15.31) and (15.19) (which latter gives X_Q) will show that it corresponds exactly with $T\sigma$. It might be added that while this method is hardly rigorous enough in the present case to serve as a means for calculating $T\sigma$, and so σ, it can be used in certain simple isothermal processes to calculate the entropy production. For instance, where an electric current I is flowing in a conductor under a potential difference E, the rate of working of the electrical supply (alternatively the energy dissipation ($T\sigma$) in the seat of the irreversible process, the conductor) is EI. Consequently the entropy production in the (isothermal) conductor is $\sigma = \dfrac{EI}{T}$, and if I is taken as the flux the thermodynamic force is $\dfrac{E}{T}$. A similar simple calculation gives the entropy production associated with the flow of a viscous liquid down a pipe. For an isothermal chemical or diffusion process the energy dissipation is simply the free energy decrease per unit time, and division by T leads directly to σ.

The Energy Requirement of Active Transport

It has been a matter of controversy whether to any appreciable extent active transport of water can contribute to the water balance of cells. This can be discussed with reference to the steady state which supervenes when water transport has ceased and the turgor pressure is a maximum. If it is assumed for simplicity that the cell sap has zero osmotic potential the question then becomes, what energy dissipation ($T\sigma$) is associated with a turgor pressure (ΔP) when the process of water intake has ceased ($J_m = 0$)?

The general expression for the energy dissipation in a system with two fluxes[1] is

$$T\sigma = T(J_1 X_1 + J_2 X_2). \tag{15.43}$$

If the subscript in $J_1 X_1$ refers to water, then at the steady state ($J_1 = 0$),

$$T\sigma = T J_2 X_2 \tag{15.44}$$

with

$$J_1 = L_{11} X_1 + L_{12} X_2 = 0 \tag{15.45}$$

$$J_2 = L_{21} X_1 + L_{22} X_2. \tag{15.46}$$

From equation (15.45)

$$X_2 = -\frac{L_{11}}{L_{12}} X_1 \tag{15.47}$$

[1] We are not now limiting the consideration to thermo-osmosis.

and using this and equation (15.46) in (15.44) with Onsager's relation gives

$$T\sigma = TL_{11}X_1^2 \left[\frac{L_{11}L_{22}}{L_{12}^2} - 1\right]. \tag{15.48}$$

This is as far as the theory can take us. On account of the inequality corresponding to equation (15.42) the expression in the square brackets is always positive, but beyond that it is not possible to say how large or how small it is. If the cross-coefficients $L_{12} = L_{21}$ are large (there being what might be called a well-marked active process) this will reduce the energy dissipation corresponding to a given value of X_1, and for the case we worked out this force is proportional to ΔP. On the other hand, if L_{11} or L_{22} is large this will put up the necessary energy expenditure especially in the case of L_{11}, which appears twice. The coefficient L_{11} is proportional to the ordinary permeability of the membrane to water; in fact it only requires two or three small modifications such as dividing by the area to turn it into the permeability. The coefficient L_{22} is a rough sort of thermal conductivity in the case of thermo-osmosis; but were the second flux J_2 to be one of electrolytes in an electrokinetic system then it might be described loosely as the 'ordinary' permeability to these. It is thus true to say that where the 'ordinary' permeabilities of a cell membrane are high the energy expenditure required to actively maintain a turgor pressure is likely to be high as well, but that this can be offset to an undefined extent by high cross-coefficients. Incidentally, having raised the subject, electrokinetic phenomena represent a type of transport process where cross-coefficients are appreciable, though this is merely a sophisticated way of saying that the phenomena are not at all difficult to see.

Summing up therefore, it can be said that the energy dissipation in actively-maintained turgor is likely to be high where the cell permeability, especially to water, is high; that it increases with the *square* of the pressure set up (a point not noticed before); but that so far as thermodynamics is concerned it has a lower limit of zero. Additional information is needed before its magnitude can be fixed more closely than this.

Stationary States and Minimum Entropy Production

The fluxes that occur in a system not in equilibrium are naturally in such a direction that they tend to relieve the forces producing them. If the system is left to itself the forces will therefore eventually all become zero, and the fluxes likewise. However, it sometimes happens that one or more of the forces are artificially (so far as the system is concerned) maintained at a finite value, the simplest case being when a single force is maintained at a fixed value.

For simplicity attention will be limited to this particular case, and it will be discussed with reference to the thermo-osmotic system investigated above. Where the principles which emerge are general ones applicable also to larger and more complex systems this fact will be pointed out.

The equations which describe the movement towards equilibrium of a thermo-osmotic system are ((15.32) and (15.33))

$$J_m = L_{mm}\frac{v\Delta P}{T} + L_{mQ}\frac{\Delta T}{T^2} \qquad (15.49)$$

$$J_Q = L_{Qm}\frac{v\Delta P}{T} + L_{QQ}\frac{\Delta T}{T^2}. \qquad (15.50)$$

These equations make no stipulations about the sort or size of divisional membrane present, the mechanism by which water crosses it, or the way the surroundings are reacting on the system. It can therefore be assumed that experimental arrangements may be made to hold ΔT at a fixed value. What are the consequences of this?

To begin with (since a start must be made somewhere) suppose that the system is at equilibrium ($\Delta P = 0$, $\Delta T = 0$), and that then the temperature difference ΔT is applied. Writing $\Delta P = 0$ in equations (15.49) and (15.50) it can be seen that fluxes in both heat and matter will begin, given by

$$J_m = L_{mQ}\frac{\Delta T}{T^2} \qquad (15.51)$$

$$J_Q = L_{QQ}\frac{\Delta T}{T^2}. \qquad (15.52)$$

Since ΔT is to be maintained constant heat will have to be added to one side and removed from the other; further the flux of matter will inevitably destroy the equality of pressure. A differential ΔP will arise, directed in such a way (see the negative sign in equation (15.39)) as to progressively reduce the flux of water (J_m). The differential ΔP will grow as water moves across until it reaches, asymptotically, such a value [1] that the flux J_m becomes zero. This will signify the stationary or steady state. Meanwhile the rise of ΔP has not been without influence on the flow of heat (J_Q). On account of the Onsager relation $L_{Qm} = L_{mQ}$ it follows that ΔP is directed in such a way as to reduce J_Q as well as J_m; but the most interesting fact is that of all the values which ΔP might have assumed the one at which it eventually settles is that which makes *the energy dissipation, or entropy production, a minimum*. This is very easily proved.

---

[1] Which has been called above the thermo-osmotic pressure.

First of all, from equation (15.49), the condition for the steady state is
$$J_m = L_{mm}X_m + L_{mQ}X_Q = 0 \tag{15.53}$$
where the forces have been written as X's for short. But the entropy production (cf. equation (15.41)) is
$$\sigma = L_{mm}X_m^2 + (L_{mQ}X_mX_Q + L_{Qm}X_QX_m) + L_{QQ}X_Q^2. \tag{15.54}$$
With a fixed value of X_Q the minimum value of σ is found by equating $\dfrac{\partial \sigma}{\partial X_m}$ to zero. This gives
$$2L_{mm}X_m + (L_{mQ} + L_{Qm})X_Q = 0 \tag{15.55}$$
as the condition for the minimum. On account of the Onsager relations equation (15.55) is clearly the same as equation (15.53).

What has been proved for a simple system with only two fluxes is in fact a principle of almost[1] universal generality, applying to all sorts of irreversible processes in systems of all degrees of complexity. It can be stated broadly like this. When a system whose state of disequilibrium described by a number of independent forces $X_1, X_2, X_3, \ldots$ is maintained with one group (A) of these forces constant, the system changes in such a way that all the fluxes corresponding to the other group (B) progressively become zero. When this stationary state has been reached the forces in group B will have assumed such values that the entropy production, or energy dissipation, of the system is a minimum in the circumstances (circumstances constituted, of course, by the set of values of the forces in group A). It may further be remarked that the stationary state is described as being of an *order* equal to the number of forces in group A. The case just discussed is thus a stationary state of the first order; further, true thermodynamic equilibrium may be called a stationary state of zero order.

Further Remarks

If the reader can tolerate flippancy for a moment it may be remarked that the principle that has been discussed is exemplified[2] very well by the behaviour of bath water. At the moment when the plug is pulled out there begins an efflux of water under what may be considered a fairly steady pressure head. This flux initiates others; in particular the sponge and flannel move to the exit and proceed to plug the hole as effectively as they can. When their motion has ceased ($J_{\text{sponge}} = $ zero) the bath is emptying at the minimum rate possible in the circumstances, and the energy dissipation is at its lowest. Any further movement of the sponge—unfortunately it sometimes breaks

---

[1] The exceptions are well understood and need not be of concern here. They occur principally when external magnetic fields are present.

[2] The analogy that follows should be viewed in the light of the remarks on p. 112.

up and disappears—would result in a rise in the dissipation. The analogy may not be an exact one, but it is instructive.

In a more serious vein it can be remarked that when a system, initially in internal equilibrium, is subjected to a 'force' (such as a temperature difference) its own entropy S may fall as it moves to the steady state. This happens for instance when a temperature gradient is applied to a solution. The so-called Soret effect follows, and a concentration gradient develops. This gradient might be described as a very rudimentary structure in what would otherwise have been a homogeneous phase. Of course, the reduction in entropy which it represents is more than counterbalanced by changes in the surroundings which impose the original constraint; but it is significant to a biologist all the same. This point is raised again in the Epilogue.

A case of differing interest to biologists is provided by what are called isoaffine chemical reactions. Consider an open system in which the following system of chemical reactions takes place:

$$\xrightarrow{\text{transport process}} A+B \rightleftharpoons M \rightleftharpoons Y+Z \xrightarrow{\text{transport process}}$$
$$\text{\quad\quad\quad\quad\quad\quad} \updownarrow$$
$$\text{\quad\quad\quad\quad\quad\quad} N$$

Imagine that substances A and B are supplied in such a way as to maintain their chemical potentials constant; similarly that Y and Z are removed in the same fashion. This will mean that the affinity of the main reaction sequence is kept at a steady value (hence the name 'isoaffine'). It can be shown that the Onsager relations apply to such a system, though it is not a closed one like the thermo-osmotic example that was discussed. If there is coupling between the main reaction sequence and the side reaction (M⇌N) (that is, if the cross-coefficients are not zero) then the progress of the main reaction will displace the side reaction from its point of true thermodynamic equilibrium; and this displacement will react on the main sequence to reduce its rate to the minimum value possible. It is not difficult to see a possible relevance of this to living organisms. The reactants A and B may be food and oxygen; and the products Y and Z carbon dioxide and water. The reaction M⇌N will appear to go in the wrong direction (until it comes to rest), like an active transport against a concentration gradient; in fact, it will be an 'active' chemical reaction. Further, if when the steady state has arrived the chemical potential of N is artificially altered this will be reflected in an *increase* (from the minimum position) of the energy dissipation of the isoaffine reaction chain. This may perhaps be a clue to some aspects of such phenomena as salt respiration in plants, where the addition of salts to a steady system promotes an increase in the oxygen uptake[1].

[1] It would be a confirmation of this interpretation if *removal* of salts from the medium had the same effect.

The Efficiency of the Metabolic Pattern

The principle of minimum entropy production has an important bearing on the overall metabolic pattern of living systems. From the discussions in previous chapters it will be appreciated that an energy transaction is accomplished with greater thermodynamic efficiency the more nearly it is carried out reversibly. This ideal is approached when all out-of-balance forces are kept small. Frictional forces (in solid systems) impose a lower limit on these out-of-balance forces; they must at least be equal to the friction or no movement takes place. Friction therefore must be reduced in such systems. In chemical systems the activation energy[1] plays the role of frictional forces, setting a sort of minimum effort to be exercised for the reaction to take place at all. Thermodynamically the activation energy represents waste (see Fig. 15.4). Thus it is an obvious wisdom, teleologically speaking,

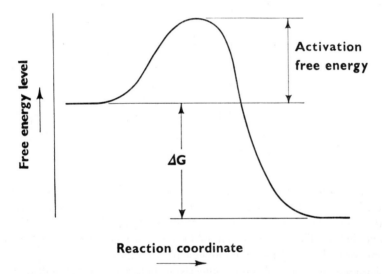

Fig. 15.4. The activation free energy and the Gibbs free energy of a chemical reaction.

to lower the activation energy by enzymes quite apart from the accompanying improvement in rates. However, even if the activation energy is reduced to zero, $-\Delta G$ remains as an out-of-balance force. It is here that the present principle comes in. If circumstances exist for an overall reaction to proceed in stages of small $-\Delta G$ (that is, if the necessary enzymes and auxiliary substances are present) then the minimum dissipation principle would seem to ensure that the overall reaction is in fact channelled along such a course.

[1] Or better, activation free energy.

This is not a conclusion at once intuitively obvious, so perhaps a simple and somewhat artificial analogy[1] will help. Imagine a tank (Fig. 15.5) with several resistive outlets ending in vertical pipes of differing lengths. Water will flow from all the outlets but at unequal velocities because the various outlets impose different suctions. The several exits represent different biochemical pathways; to be more concrete the water can be imagined to represent ATP, with the longest exit standing for hydrolysis to ADP and inorganic phosphate,

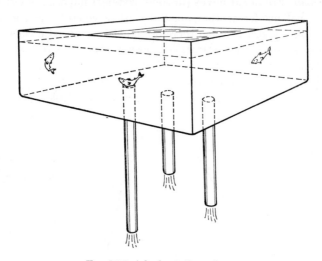

Fig. 15.5. A hydrostatic analogy.

while the others represent breakdown to ADP and sugar and alcohol phosphates. We imagine the tank to contain fish as well, and that these are frictionally affected by the flow of water. Of course they will tend to be carried to the exit holes[2]. Their own powers of locomotion (the best analogy we can provide for molecular Brownian movement) will offset this tendency, and we may suppose that the holes most effectively plugged with fish will be those with the most powerful suction; that is, the ones which represent the stages in which ATP breakdown has the greatest $-\Delta G$. The net result will be that after a short initial period the outflow of water will be preferentially directed into the pipes with low suctions. This analogy may not be a perfect one, but it does help to show how it may come about that living systems are so marvellously organized for conducting their metabolism in a way reasonably close to the thermodynamic ideal of reversibility. Of course, the principle

---

[1] See footnote 2, p. 253.

[2] The rougher their skins the greater will be the 'cross-coefficients' and the lower the value of $T\sigma_{min.}$ attained (cf. equation (15.48)).

of minimum dissipation does not account for everything; the fish must be provided, the holes must not be too large, nor the suction too great, or all the fish will disappear (what happens in high temperature combustion is a parallel to this).

The Organization of Metabolism for Thermodynamic Efficiency

This would seem to be an appropriate place to discuss a matter of great theoretical and practical interest, how living systems organize their overall metabolic pattern so as to achieve the highest thermodynamic efficiency. This is a large subject with many ramifications, and all that can be done is to discuss certain guiding principles. These principles will be applicable to many different types of change; in what follows sometimes we shall relate them to metabolic reactions, and sometimes (by way of variety or clarity) to transport processes. But in themselves they are quite general.

Passive Processes

Consider firstly the following two cellular reactions[1]:

(i) phosphoglyceric acid → pyruvic acid + $PO(OH)_3$ (inorganic phosphoric acid)

$$\begin{array}{c} CH_2OH \\ | \\ C{\scriptstyle\diagdown}^{H}_{O-PO(OH)_2} \\ | \\ COOH \end{array} \quad \rightarrow \quad \begin{array}{c} CH_3 \\ | \\ CO \\ | \\ COOH \end{array} \quad + \quad PO(OH)_3$$

(ii) $PO(OH)_3$ (inorganic phosphoric acid) + ADP → H_2O + ATP

The first of these is highly exergonic[2]; that is, it has a large negative $\varDelta G$. The second on the other hand is endergonic, having a numerically rather smaller and positive $\varDelta G$. Thermodynamically therefore the first reaction could be linked with the second to give a spontaneous overall process in the stoichiometrical amounts indicated. In actual fact the linkage occurs through

---

[1] It is not implied that these two reactions actually occur together in the cell as written, but only that the outcome is 'as if' they occurred. They might be called 'virtual' reactions, as we speak of a 'virtual' image in optics. The actual course of the metabolism seems to be as indicated by the later sequence (iii).

[2] At typical concentration levels.

a third substance, phosphoenolpyruvic acid, and the 'real' metabolic pathway (as opposed to the 'virtual' one) can be written:

(iii)
$$\begin{array}{c} \text{phosphoglyceric acid} \\ \downarrow -H_2O \\ \text{phosphoenolpyruvic acid} \\ \downarrow +ADP \\ \text{pyruvic acid } +ATP \end{array}$$

The two stages of this sequence are *each separately exergonic*, and it is this fact which constitutes the main reason for regarding the synthesis of ATP which it entails, as a passive process. In other words, it is not essential to look for any active mechanism, though that is not to say that none is present. In passing it may be noted that, provided the concentration of the intermediate is steady, the production of ATP is stoichiometrically related to the consumption of phosphoglyceric acid, and as will be seen later this cannot be the characteristic of an active process.

It is a fairly simple matter to discuss the energy economy of passive processes. Thermodynamically the greatest economy is achieved when the whole process occurs in a manner as close to reversibility as possible. This means that the ΔG of each stage must have a negative value which is as numerically small as is practicable. In theory *any* chemical reaction can comply with this requirement; even if its *standard* affinity is numerically very large, its *working* affinity can always be made small and be given a negative sign by adjustment of the various concentrations, as comparison of equations (11.28) and (11.30) shows. However in practice, if a metabolic sequence is made up of stages with large *standard* affinities (for argument's sake all assumed positive) operating such a sequence so that each stage entails only a small *working* affinity (or free energy decrease) will mean a vast change in concentration as between the first reactants and the ultimate products. This should be fairly clear, but to emphasize the point consider the simplest possible reaction chain:

$$A \rightarrow B \rightarrow C$$

where a plain isomeric change occurs in two stages. If each stage is highly exergonic (under standard conditions) the equilibrium constants $K_1 = \dfrac{m_B}{m_A}$ and $K_2 = \dfrac{m_C}{m_B}$ will be high (see equation (11.30)); suppose each of them is taken to be 10,000. Consequently if the reaction sequence is traversed in a manner not too far removed from thermodynamic reversibility the working concentration of B must be roughly 10,000 times that of A, and the concentration of C roughly 10,000 times that of B. Thus C will be present as a concentration of nearly 100,000,000 times that of A. Since the upper limit of any concentration is clearly set by solubility and other factors the actual concentration of A must needs be exceedingly low, and this will impose a severe limit on the rate of the overall reaction, a very important qualification as we shall see. Thus it follows that for efficiency the metabolism must be organized in steps whose *standard* affinities are not too great.

The Question of Rate

At this point two further considerations arise. In the first place, working close to thermodynamic reversibility means working slowly, since all out-of-balance forces are kept low. Living things, however, have to make their way against competition from other living things, and they also have to complete their life cycles within time limits imposed on them by matters outside their control, such as the seasons. These considerations put a premium on rate, for an organism which can grow or move more quickly may on that account be more successful than another which nevertheless uses its energy resources more efficiently. It is only up to a point therefore that it pays to improve efficiency in the sense we are discussing; beyond that point the slowing up of the whole tempo of life is a severer handicap than the failure to realize fully the available free energy resources.

Secondly, there is the question of the stability of the metabolic pattern. Where a process takes place close to reversibility it does not take much change in the environmental situation to alter its rate appreciably, and perhaps even to reverse its direction[1]. Plants and animals live in an environment which to some extent at least is subject to erratic and inescapable variations, and it is clearly undesirable that their metabolism should be too much at the mercy of these. Changes in temperature, water stress or nutrient supply are the sort of things we are thinking of. Stability is improved when metabolic steps take place under a reasonable affinity; that is, where a fair element of thermodynamic irreversibility is involved. In fact the situation is analogous to human practice where an operation is often conducted less efficiently for the express purpose of securing greater steadiness (one can think of the resistance inserted in a battery-charging circuit, or of the light air-vane which stabilizes the running of the clockwork mechanism in a striking clock).

---

[1] When $-\Delta G$ in equation (11.28) is small the two terms on the right hand side are nearly equal and their difference becomes very sensitive proportionally to a change in one or other of the concentrations.

Active Processes

Now consider the question of active mechanisms. To fix our ideas consider a system where two independent things can happen; that is, where two fluxes J_1 and J_2 can be referred to. These might represent the transport of ions and water, or they might equally well represent two independent[1] chemical reactions, the second of which resulted say in the synthesis of ATP (in fact the very same two reactions we have considered above might be suitable). Suppose the first reaction proceeds under an affinity[2] X_1 kept constant by the continuous provision of 'reactants' and removal of 'products'. Concretely, we may think of salts provided by the soil solution to the root and removed by xylem transport, or of carbohydrates and oxygen continuously supplied to other tissues, with carbon dioxide and water as continuously removed. Now write for the two reactions

$$J_1 = L_{11}X_1 + L_{12}X_2 \qquad (15.56)$$
$$J_2 = L_{21}X_1 + L_{22}X_2 \qquad (15.57)$$

and consider the simple situation where the primary reaction (15.56) drives the other (15.57) to its stationary position and holds it there. Then $J_2 = 0$ and from equations (15.57) and (15.56):

$$X_2 = -\frac{L_{21}}{L_{22}}X_1 \qquad (15.58)$$

and

$$J_1 = \left(L_{11} - \frac{L_{12}L_{21}}{L_{22}}\right)X_1. \qquad (15.59)$$

To refer again to the examples, the situation now is that a continuous flux (J_1) of ions is maintaining water at an increment of pressure (measured by X_2); or a continuous consumption of carbohydrate and oxygen is maintaining ATP at an enhanced level (again measured by X_2).

In passing, it can be noted that the existence of the cross coefficients $L_{12} = L_{21}$ implies an effect somewhat analogous to friction; for equation (15.59) shows that the flux J_1 is less than it would be were these coefficients zero (when it would be simply $J_1 = L_{11}X_1$).

Now it is obvious that the active effect being discussed has done something potentially useful; it has, in the examples, boosted water or ATP to a

---

[1] 'Independent' here means that the degrees of advancement (ξ_1 and ξ_2) of the two reactions can each be specified arbitrarily.

[2] The 'affinity' here is the Onsager force and is $\frac{1}{T}$ times the ordinary affinity A in the case of the chemical example.

higher chemical potential, and this useful effect is measured by X_2 (equation (15.58) gives the *maximum possible* effect). The cost of this useful result is in turn measured by the steady energy dissipation[1] $T\sigma = TJ_1X_1$ needed to maintain it, where X_1 is fixed by hypothesis and J_1 has the value given by equation (15.59). Since X_2 is proportional to X_1 (see (15.18)) there is this result: the useful effect of an active mechanism is proportional to the out-of-balance force X_1 under which the driving reaction proceeds, whereas the cost is proportional to the square[2] of X_1. This means that active mechanisms are most efficient when the driving force X_1 is low; in fact the useful effect per unit cost is halved when X_1 is doubled and *pro rata*.

Some Further Remarks on Active Mechanisms

It was suggested above that if a concrete example is required of one chemical reaction being actively influenced by another then perhaps even the passive sequence (iii) (p. 258) would do. In a *purely speculative* sense this suggestion can be justified as follows. When the first step takes place, the product (phosphoenolpyruvic acid) is formed as a population of molecules with a 'temperature' slightly different from the ambient. How this happens is discussed in a moment, but here it need only be remarked that if the second step follows the first quickly and closely enough it will be influenced by this difference in 'temperature', and in addition to the ordinary affinity of this step it will experience a 'push' (perhaps a 'pull') related to this difference. This is an active effect, for the 'proper function' of a temperature difference is to promote flow of heat[3], and it will exhaust itself in doing so. The promotion of chemical change in the way described is thus an example of a cross effect. Provided that the ATP resulting from the *passive* force of ordinary affinity is not consumed by a further reaction, then its concentration is maintained without further consumption of energy; but when the ATP level is *boosted by the cross effect* it can only be held constant by the continuous dissipation of energy, for the scalar temperature difference which boosted it must be kept up. Again, it is this necessity which justifies us in regarding the mechanism as active.

The discussion of the metabolic sequence (iii) suggests two further observations. If the first stage in the sequence is poisoned by an enzyme inhibitor the production of ATP will cease. This sensitivity to inhibitors is clearly consistent with the whole sequence being a purely passive one, 'downhill' all the way. With this consideration in mind it should also be clear that the

---

[1] It is this steady energy dissipation which justifies the epithet 'active'.

[2] The cost being $TJ_1X_1 = T\left(L_{11} - \dfrac{L_{12}L_{11}}{L_{22}}\right)X_1^2$.

[3] This is spoken of in the present case as a *scalar* flow of heat, for it has no direction in space.

finding that salt uptake by cells is prevented or slowed down by respiratory inhibitors *does not necessarily mean that the uptake is active*. Until it has definitely been shown how the connection between salt uptake and respiration is effected it may be better therefore to speak of it as a metabolic uptake rather than an active one. The two descriptions are not synonymous.

Again in connection with metabolic uptake of ions the attempt has sometimes been made to show that it is stoichiometrically linked with some component of the respiration, a definite number of ions being concerned for each oxygen absorbed. This suggests a second observation: *an active linkage cannot be a stoichiometrical one*. This is easily proved, for if the case of two fluxes linked actively (that is, with significant cross-coefficients) is considered, then as before

$$J_1 = L_{11}X_1 + L_{12}X_2$$
$$J_2 = L_{21}X_1 + L_{22}X_2.$$

However, since the forces X_1 and X_2 are independent (that is, fixing the value of one arbitrarily leaves us free to fix the value of the other arbitrarily as well) the ratio of J_1 and J_2 cannot have any definite value. As an example, in the case of an electro-osmotic system (which provides a good example of an active mechanism), when the pressure difference across the membrane is low the water transported per unit amount of ions is large; but as the pressure difference builds up it becomes less and less, eventually dropping to zero. The ion movement may of course continue. Thus the demonstration of a fixed ratio between two fluxes mean that there cannot be an active linkage between them.

Cellular Membranes and Activation Energies

Many cellular membranes appear to be partly lipoidal in nature. This means that they constitute a considerable potential energy barrier to the passage of ordinary hydrophilic metabolites, a fact evidenced by the high Q_{10} of penetration. As a consequence they exert a kind of filtering action on the solute molecules impinging on them, allowing the 'hotter'[1] ones to pass but turning back the 'colder' ones. Thus it comes about that when a hydrophilic solute diffuses across such a membrane its molecules immediately on the far side constitute a population with a 'temperature' greater than the ambient, and therefore chemically more influential. If it is possible for them to enter a chemical reaction before their higher temperature has been equalized (admittedly an extremely rapid process) use may be made of their heat energy to drive the reaction away from the rest-point at which it would otherwise settle. In this way the free energy of diffusion might drive a chemical

---

[1] These adjectives will be readily understood in the context.

synthesis; but the linkage clearly depends on the provision firstly, of membranes constituting potential energy barriers, and secondly, of reaction spaces extremely close to the membrane. It is perhaps possible to view the internal subdivision of mitochondria and chloroplasts at least partly in this light.

What has been said about membrane action holds also for chemical reactions, the chemical activation energy taking the place of the energy of penetration, and allowing only the 'hotter' molecules to pass from one side to the other of the reaction equation. Similarly, a second chemical reaction will only be influenced by the higher temperature of the products of the first if its site is in extremely close proximity. This constitutes one reason, perhaps, for the structural organization of enzyme systems rather than for their existence in free solution.

It will probably be obvious to the reader that the activation energy of penetration or chemical reaction is linked very closely with the heat of transfer Q^* of the transport or chemical process. Further, the heat of transfer is dependent on significant Onsager cross-coefficients (cf. equation (15.37)). What has been discussed therefore is how the ultrastructural features of internal membranes and spatially organized enzymes can influence the cross-coefficients. In the light of all this it will be clear that such architectural features of the cell may have a marked effect on its energy economy.

Heat Engines Again

It was seen that in the case of the photochemical stage in photosynthesis (another active process), the leaf functions as a heat engine. In the hypothesized active processes just discussed and in which the heat of transfer Q^* plays a part the living system also functions as a heat engine. The occurrence of an active mechanism means that part of the free energy decrease of the primary reaction (a decrease which might otherwise be considered as lost) has been salvaged in a useful form. It was seen earlier that living things are under the obligation of not conducting their metabolic processes too slowly. This means that reactions must proceed under a reasonable affinity, or to put it in other words, with a fair element of thermodynamic irreversibility. The level of affinity required depends on the activation energies of the processes, which inversely affect the coefficients L, so that a reaction with a high activation energy needs to proceed under a high affinity, or its rate will be too slow. However, high activation energies also mean high heats of transfer, and this enhances the possibility of heat engine activity. Thus the very consideration that makes passive reactions wasteful to the organism (high activation energies) introduces the possibility of salvaging some of the wasted energy by active processes.

Further, this salvaging of waste heat confers an additional benefit on the organism; it tends to enhance the rate of the primary reaction and so offers the possibility of conducting it at a lower affinity without loss of speed. This situation arises as follows. Consider a reaction with a positive heat of transfer (Q^*). While this is proceeding steadily the products $\mathscr{P}$ will have a higher temperature than the reactants[1], and the temperature differential will act to slow the reaction down (compare the influence of the cross coefficient in equation (15.59)), a general effect which was noticed much earlier (Chapter 4). Suppose arrangements are made to harness this temperature differential to drive, in the active sense, a second reaction in which one or more of the hotter products $\mathscr{P}$ of the first enter as reactants. The activation energy of the new reaction will 'filter off' the hottest of the molecules $\mathscr{P}$ and in so doing lower the temperature of the remainder. Thus the primary reaction will be working against a smaller temperature differential, and will as a result proceed faster.

It may be remarked that a final and perhaps very important aspect of active mechanisms lies in their providing the organism with additional versatility and flexibility in its physiology. To pursue this subject would, however, take us too far into particularities, and probably little useful could be said in the present state of knowledge about this subject.

A Mechanical Analogy

Lest the reader should think that the suggestion that organisms might use active mechanisms to salvage some of the free energy lost in passive processes is an impossible one, a simple illustration drawn from more familiar circumstances may help to reassure him and put the matter in perspective.

When energy is transformed in a mechanical device some of it is always lost frictionally as heat. This heat results from the degradation of work, and as a consequence of this it is unlimited in the temperature at which it can appear[2]. Thus apart from practical limitations the friction of bearings may be allowed to raise them to white heat; in another context the resistor inserted in a battery charging circuit may easily be a lamp which does in fact operate at incandescence. Clearly the theoretical possibility exists of using this frictional heat to drive a heat engine, and so salvaging some at least of the energy it represents. Since its temperature has no theoretical upper limit its reconversion to work can in principle approach an efficiency of unity[3]. We only need to point out that the activation energy of a molecular process is

---

[1] The reactants suffer a temperature reduction below ambient while the products enjoy a rise above it (when Q^* is positive).

[2] This statement applies in an unqualified sense only to the *continuous* degradation of work in a *finite* system (*contra* a waterfall or a suddenly-released spring).

[3] Compare the expression $(T_1 - T_2)/T_1$ for the efficiency of an ideal heat engine.

analogous to friction, and that the harnessing of active mechanisms has an upper limit of unity (compare equation (15.42))[1] to make the parallel with metabolic systems obvious.

APPENDIX I: DIFFERENTIAL OF A QUOTIENT

If u and v are two functions of x, and $y = \dfrac{u}{v}$, the differential coefficient of y is connected with those of u and v by the formula

$$\frac{dy}{dx} = \frac{v\dfrac{du}{dx} - u\dfrac{dv}{dx}}{v^2}. \qquad (15.60)$$

In terms of differentials (multiplying through by dx) this becomes

$$dy = \frac{v\,du - u\,dv}{v^2}. \qquad (15.61)$$

Thus if the symbol Δ stands for a small enough increment then

$$\Delta\left(\frac{\mu}{T}\right) = \frac{T\Delta\mu - \mu\Delta T}{T^2}. \qquad (15.62)$$

a formula used above. Similarly it is quite easily seen that

$$\Delta\left(-\frac{1}{T}\right) = \frac{\Delta T}{T^2}. \qquad (15.63)$$

APPENDIX II: THE TEMPERATURE DEPENDENCE OF THE PERMEABILITY

It can be shown that the permeability of a membrane to water (λ) is connected with its heat of transfer by the relation

$$\frac{\partial}{\partial T}\ln\left(\frac{\lambda T}{\bar{V}_w}\right) = \frac{Q^*}{RT^2} \qquad (15.64)$$

or since $\bar{V}_w$ is nearly constant,

$$\frac{\partial}{\partial T}\ln\left(\frac{\lambda T}{\bar{V}_w}\right) = \left(\frac{1}{\lambda T}\right)\frac{\partial}{\partial T}(\lambda T) = \frac{Q^*}{RT^2}. \qquad (15.65)$$

Thus if (λT) varies proportionally very rapidly with temperature (this corresponds to a biologist's high Q_{10}) the membrane must possess a high

---

[1] Active mechanisms depend on the cross-coefficients $L_{12} = L_{21}$ and these are subject to the upper limit $\dfrac{L_{12}L_{21}}{L_{11}L_{22}} = 1$.

heat of transfer. It can be visualized that such a membrane constitutes a high potential energy barrier whose 'filtering off' action on the faster moving molecules is responsible for the high value of Q^*. In fact the value of Q^* is very largely determined by the height of the potential energy barrier interposed by the membrane. It is because a membrane with large holes interposes no such barrier that Q^* is zero and no thermo-osmotic pressure can be set up; but as in all cases entropy enters as well as energy and the potential energy barrier is not quite the whole story.

Equation (15.61) can be used to calculate Q^* when data for the permeability at different temperatures are available.

Appendix III: Active Transport—an Objection

It is possible that the reader may feel an objection to the definition of active transport given in this book on the grounds that it involves the conclusion that salt uptake does not qualify for the description 'active' merely because (i) it occurs against a thermodynamic gradient; and (ii) it is inhibited by metabolic poisons, such as dinitrophenol. In order to justify the present point of view a simple illustration is appended.

Imagine a chemical reaction

$$A + B \rightleftharpoons C \qquad (15.66)$$

in which the reactants A and B are hydrophilic, while the product C is lipophilic (perhaps an ester). This reaction will be subject to the mass-action law, and for equilibrium the equation

$$m_C = K m_A m_B \qquad (15.67)$$

will hold, where K is the equilibrium constant. Consider a system (Fig. 15.6) in which two aqueous phases (α and β) are separated by a lipoidal membrane. Suppose that m_A is ten times higher in phase β than in phase α; and conversely that m_B is ten times higher[1] in phase α than in phase β. Then by virtue of equation (15.67) both phases will be in equilibrium with the same concentration m_C of C, and since the species C alone can pass the membrane the whole system will be in transport equilibrium.

Next suppose that some of B is removed from phase β. By virtue of equation (15.67) this lowers the concentration of C in this phase, and at once C begins to diffuse across the membrane from α to β. At the interface with phase β, the reaction (15.66) proceeds in reverse, and the concentrations of both A and B in phase β begin to rise. The net result is that A and B are both transported across the membrane from α to β, and for species A *this is an uphill*

---

[1] The substance B might incidentally be present at much lower levels of concentration than substance A.

movement. Further, if the removal of B from phase β is by means of an enzyme reaction then this reaction has only to be inhibited to prevent the uptake of A.

The 'uphill' transport of A is compensated thermodynamically by the accompanying 'downhill' transport of B, incidentally in stoichiometrical amount. The whole movement is obviously a passive one, and the stoichiometrical relationship between the movements of A and B is in accordance

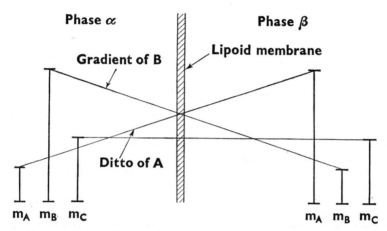

FIG. 15.6. An example to illustrate the distinction between metabolic and active transport.

with the idea that the linkage between them is not active. It might be suggested as a serious hypothesis that in the root A might stand for ions, B for oxygen, and C for a 'carrier' substance, though the model would obviously have to be developed somewhat to allow for this interpretation. It would imply that while accumulation of ions in the root interior (phase β) would be halted by anoxia, the uptake could still be a thermodynamically *passive* process.

SELECTED BIBLIOGRAPHY

Everett, D. H. (1959). 'An Introduction to the Study of Chemical Thermodynamics.' Longmans, Green & Co., Ltd., London.
  A sound and useful introduction to the subject.
Pippard, A. B. (1961). 'The Elements of Classical Thermodynamics.' Cambridge University Press, London.
  Available in paperback form this little book deals mainly with the physical applications, but has a very clear and readable account of fundamentals.
Planck, M. (1959). 'The Theory of Heat Radiation.' Dover Publications, New York; Constable & Co., Ltd., London.
  Originally published in 1912, this standard work is now available in paperback form. It is not easy, but is an essential reference for all serious students.
Glasstone, S. (1947). 'Thermodynamics for Chemists.' D. Van Nostrand Co., Inc., Princetown, New Jersey, U.S.A.
Zemansky, M. W. (1957). 'Heat and Thermodynamics.' McGraw-Hill Book Co., Inc., New York.
  Now available as a paperback this work is especially excellent on fundamentals. Contains rather more engineering and physical applications than chemical ones.
Guggenheim, E. A. (1955). 'Boltzmann's Distribution Law.' North Holland Pub. Co., Amsterdam.
  A valuable little introduction to the statistical development of thermodynamics. Expensive, and more mathematical than other titles.
Denbigh, K. G. (1951). 'The Thermodynamics of the Steady State.' Methuen, London; J. Wiley, New York.
  A useful small introduction to irreversible thermodynamics.
Setlow, R. B., and Pollard, E. C. (1962). 'Molecular Biophysics.' Addison-Wesley, Reading, Mass.; Pergamon, London.
  Contains a number of interesting and suggestive chapters.

Epilogue

Thermodynamics and Life

A question which has often been debated, and which still frequently obtrudes itself, is whether the conclusions of thermodynamics—established as they have been from arguments based on the behaviour of non-living systems—can be applied to living organisms. Living things certainly manifest phenomena, such as heredity, for which an adequate parallel can hardly be found in the realm of the non-living; but to admit that organisms go beyond the behaviour of simple physical systems in this sense is not to admit that where their behaviour is describable in physical terms it is not wholly in accordance with physical laws. What we mean is this; where the activity of the living thing really demands biological categories to describe it (such for instance as 'adaptation' or 'instinct') then there is never any question of ordinary physical laws being adequate, since as a matter of definition physical laws do not deal with biological categories. However, where the activity can be described in purely physical categories (as the uptake of ions in terms of concentration changes, oxidation of carbohydrates and so on) then there is a real question for debate. The standpoint of this book is that where a process is describable in *physical* terms, then whether it takes place in a living or a non-living system, it is subject to the same *physical* laws. In other words, in such instances organisms are subject to thermodynamic laws, and in fact to all the laws of physics and chemistry.

The justification for this attitude lies firstly in the fact that superficially it looks right, in so far as organisms certainly do not seem to be in a privileged position when it comes to their basic economy of matter and energy. If they need an element in even the minutest traces it seems to be the case that they have to be provided with it; they cannot generate it themselves. Similarly if they require to promote a change they must somehow find the necessary free energy. Secondly, it is really the only attitude that leads to a workable programme of research. Obviously in practice it is best to assume that physical laws do govern metabolism (in the widest sense of this word) and then to see how far this assumption can take us.

However, if in pursuing this policy we come up against a case in which the behaviour of a living system seems quite irreconcilable with known physical laws, it is instructive to enquire what we do then. There are in fact two courses open: either we can seek to reformulate the physical laws so that they embrace the new situation (and something of this sort happened in physics when relativity replaced Newtonian mechanics); or we can retain the laws in their original form and suppose that there is operative in the living system an essentially non-physical (i.e. metaphysical) element which is absent from the non-living systems. This is in effect what the vitalist does, and we need to remember that in doing so he is surrendering one of the basic presuppositions of all science, the assumption that without moving from the level of what is physically[1] observable, the multifarious phenomena of Nature can be shown to fit into a single coherent pattern. This creed cannot be proved *a priori*, and may therefore in the end turn out to be false; but it is supported by such a wealth of scientific experience that the great majority of scientific workers take it for granted and scarcely give it a thought. It remains, however, a creed.

Of course, living systems do pose problems of such a unique kind that thermodynamics has not yet been able to come to grips with them. Such for instance are those which arise from the fact that many metabolic reactions take place in extremely small systems, like subdivided mitochondria and chloroplasts. In such minute systems statistical fluctuations from the mean 'equilibrium' state may be of greater importance (see Chapter 9) than in the vastly larger ones which the experimenter normally encounters. As a consequence, in very small systems thermodynamic variables have to represent time-averages, whereas in ordinary systems they can be taken as having instantaneous values.

The experimenter may be familiar with this sort of situation, since in work with radioactive isotopes one can, in fact, deal with quite small populations of radioactive atoms. If a highly radioactive sample is being counted with a rate meter the needle will assume quite a steady position; but if a sample of low activity is being similarly dealt with it will show major fluctuations. A common technique in radioisotope work is the use of coincidence counting, in which the indicating instrument only registers when two events occur simultaneously. If this principle be applied to two radioactive samples the fluctuations of the final 'coincidence' record will clearly be much coarser than would be the records for each sample separately; very rapid fluctuations have by the coincidence principle been replaced by something much more drawn-out in time. It is interesting to speculate whether possibly something like this might occur in cell organelles, the 'outputs' of different mitochondrial compartments say, being linked so that their statistical fluctuations become slow enough to be of significance on the time scale of grosser physiology. The 'flicker' phenomenon in red blood cells may possibly be an example of this sort of thing on a fairly fine scale.

The Arrow of Time

Alone among all the great generalizations of physics the Second Law of Thermodynamics distinguishes between a forward and a backward direction

---

[1] That is, observable directly by the physical senses or indirectly by means of instruments.

of time. Outside of physics, however, there are at least two other principles which place an arrow on time; the first is Memory, and the second is the Principle of Evolution. It would seem very unlikely that the Principles of Memory and of Evolution are derivable from the Second Law, and this raises the interesting question, which we cannot now pursue, as to whether the three principles would always give the same answer to the question, which of two past instants came before the other. In practice the question would be very difficult to settle. Reliable physical clocks can easily be obtained, the test of their reliability being that they satisfactorily corroborate one another. However it is much more difficult to satisfy this criterion with memories; and in the case of the evolutionary 'clock', the judgement as to whether one form of an organism was better adapted than another related to it in time (this being one way in which the evolutionary clock is 'read') is too subjective to be of the required precision. However, the relationship between the three principles remains an interesting problem.

In one related respect a more positive answer can be given. The Second Law might be called a 'downhill' principle, since it is concerned with increase in entropy or disorder. The Principle of Evolution, however, might with equal justification be called an 'uphill' principle, since it is concerned with adaptation or improvement. On this account the two principles have sometimes been held to be in conflict; and granted the validity of the Second Law it has been maintained that the upward trend of living things requires more than natural agencies to explain it. This objection has more weight when applied to ontogeny than to phylogeny; but even here it seems to be invalid. It was mentioned in Chapter 15 that when a temperature difference is applied to the opposite sides of a vessel containing a solution (a thermal diffusion cell) a rudimentary structure develops in the form of concentration gradients; and that in the context of an overall increase of entropy of the system plus surroundings, the entropy of the system itself may decrease. It is not difficult to argue from this simple case (which of course presents no difficulty) to the conclusion that the growth of an organism is quite in harmony with the Second Law. All that the law requires is that when an organism grows there should be an overall increase of entropy of the organism plus its environment; order may increase in the organism provided the environment increases still more in disorder. Although detailed measurements would be extremely difficult to make, it is not hard to believe that this condition is always met.

An Ultimate Problem

The Second Law of Thermodynamics does however pose some difficulties of a rather different and more profound kind. Life as we can observe it is always associated with matter and energy, and matter and energy are the two entities with which the Second Law is concerned. It seems therefore as if life itself must inevitably be involved in the universal process of degradation of which the law speaks. Every aspect of physical life, after all, turns on the existence of a lack of equilibrium; for instance, the ability to see presupposes non-equilibrium between the radiation which the eye (like all material bodies) is emitting and the radiation entering it from outside. If the universe, or at least the solar system, is moving inexorably towards thermodynamic equilibrium then so far as the biologist (or psychologist) is concerned that means death.

It is at this point however that we need carefully to avoid falling into a very easy logical pitfall. The Second Law, in respect of which the question at issue appears in its most acute form, has been established on the basis of physical observations; consequently its range of validity is limited strictly to what can be physically observed. We have no *a priori* right to apply it for instance to entities such as mind. We need to remember that in a very real sense what the analytical biologist studies is not life, but only the physical system which mediates it. Life is an element which transcends anything, in fact, which the scientist as such can investigate. Always and inevitably there is the investigating mind peering into things, but itself out of sight. Thus to suppose that even the Second Law, which has been called the most absolute of all physical laws, imposes limitations on the operations of mind is to make an entirely unwarranted extrapolation from the physical facts. The only legitimate conclusion from these relates to the physical universe, and to the physical systems within it which mediate one of its most remarkable phenomena, life. Once this is realized one profoundly human problem raised by the Second Law vanishes.

> Of old hast Thou laid the foundations of the earth;
> And the heavens are the work of Thy hands.
> They shall perish, but Thou shalt endure:
> Yea, all of them shall wax old like a garment;
> As a vesture shalt Thou change them and they shall be changed:
> But Thou art the same,
> And Thy years shall have no end.
> The children of Thy servants shall continue
> And their seed shall be established before Thee.
>
> Psalm 102

Subject Index

A

A (affinity), *see* Affinity
Absolute temperature, 50, 131
Accumulation ratio
  effect of non-ideality on, 187-188
  variation of with concentration, 184-185
  variation of with valency, 185-186
Activation energy, 255, 262-263
  and friction, 255, 264
Active mechanisms, inhibitors as tests for, 261-262, 266-267
Active processes, 5-6, 248-249, 254, 260-262, 263, *see also* Active transport
Active transport, 4, 100-101, 203, 248-249, 266-267
  energy requirement of, 250-251
Activity
  absolute, 176
  relative, 134-135, 176-177
Activity coefficient
  formula for, 179, 188
  ionic, 176-177
  mean, for salts, 178, 179-180
Adenosine triphosphate (ATP), 228-229
Adiabatic, 74, 121
Affinity (A)
  of chemical reaction, 143-144
  of transport process, 169-171
  relation to free energy change, 162
  standard, 160-161, 171
Affinity, Onsager, 235
  of chemical reaction, 260
Ambient pressure and osmotic potential, 88
Analogies, mechanical, 112
Angle, solid, 229-230
Avogadro's number (N), xii

B

Bacteria, 115
Bath-water analogy, 112, 253-254
Black-body radiation, 217, 223
  Planck's formula for, 230-231
Boltzmann's constant (k), xii
Brownian movement, 7

C

Calorimeters, constant volume and pressure, 107
Carnot cycle, 73
Change, measure of, 62 *et seq.*
Chemical
  affinity, 143, 144, *see also* Affinity
  energy, 12, 13
  equilibrium constant, 159-160
  reversibility, 48
  work, 13
Chemical potential, 130 *et seq.*
  alternative expressions for, 134
  and free energy, 131-134
  dependence of on composition, 147-150
  dependence of on pressure, 152-155
  dependence of on temperature, 157
  of salts, 177-179
  standard, 149
Chemical potentials of components and total free energy, 136
Chemical reactions
  free energy change of, 158-161
  independent, 260
  isoaffine, 254
  reversible, 48
Chloroplast structure, 263
Closed system, 14
Coefficient
  activity, *see* Activity coefficient
  partition, 169-170
Composition, description of, 145-147
Conservation of Energy, Principle of, 10 *et seq.*
Cross-coefficients, Onsager, 248-249, 251, 256, 263
Cross-differentiation identity, 35

D

Degrees of freedom, molecular, 126
Differential
  of a quotient, 265
  total, 29-31
Differential coefficients, partial, 26 *et seq.*

Diffusion potential, 174
Diffusion-pressure deficit (D.P.D.), 4, 206-207, 209
   inadequacy of term, 206, 209
Dilution, free energy change of, 147
Dissipation of Energy
   in irreversible processes, 249-250
   Principle of, 79
Donnan equilibrium, 94, 180-182
   electrical potential in, 183, 186-187
   influence of pressure on, 191-192
   osmotic unbalance in, 189
Donnan strength, relative, 184
Dynamic equilibrium, 98

E

Economy, energy, 257 et seq.
Efficiency
   'economic', 213, 225-226
   of metabolic pattern, 255-259, 263-265
   quantum, 214, 226-227
   thermodynamic, 213
Electrical potential between phases, 173-175
Electrochemical potential of ions, 175-176
   effect of pressure on, 191
Electrode
   glass, 198-199, 201-202
   redox, 199-200
   standard, 196-199
Electrodes
   general theory of, 192-194
   reversible, 196-197
Electrokinetic phenomena, 251
Endergonic process, 109
Energy, 9 et seq.
   chemical, 12, 13
   dissipation of, 79, 249-250
   economy, 257 et seq.
   free, see Free energy
   internal, 16, 17, 53
   kinetic, 13
   potential, 13
   Principle of Equipartition of, 126
   thermal, 14, 51 et seq., 125 et seq.
Enthalpy (H), 17, 107
Entropy (S), 62 et seq., 80 et seq.
   and disorder, 120, 125
   and probability, 116
   change on melting, 125-127
   factors influencing, 119-120
   formal derivation of, 67 et seq.
   Principle of Increase of, 82
   of perfect gas, 122-125
   of radiation, 222-224
   partial molar, 157
   production of in irreversible processes, 238, 248
Equations, addition of thermodynamic, 162 et seq., 171-172
Equilibrium, 97 et seq.
   absence of barriers to, 102
   analytical conditions for, 103, 105
   conditions for chemical, 141-142
      transport, 134
   Donnan, see Donnan equilibrium
   dynamic, 98
   general equations for in closed system, 105, 107
   in open system, 137-138
   processes concerned with, 101
   types of, 98 et seq.
Equilibrium constant
   chemical, 159-160
   transport, 169-170
Equipartition of Energy, Principle of, 126
Evolution, Principle of, 271
Exergonic process, 109, 229
Extensive properties, 19, 78, 136
Extent of reaction (ξ), 142, 143, 171

F

F (Helmholtz free energy), 85 see also Free Energy
Faraday, the ($\mathscr{F}$), xii
Fick's law, 234
First Law of Thermodynamics, 11 et seq.
   alternative statements of, 19
Flow of heat, scalar, 261
Free energy, 81 et seq.
   and chemical potential, 131-134
   and direction of spontaneous change, 109-110
   and equilibrium, 108
   direct establishment of idea, 86
   Gibbs, 85
   Helmholtz, 85
   partial molar, 37-38, 136
   statistical implications of, 160

SUBJECT INDEX

Free energy change
  and affinity, 162
  influence of temperature on, 168-169
  in respiration, 165-168
  of chemical reaction, 158-161
  of dilution, 147
  of transport process, 169-171
  standard, 161
  tables of, 165, 166
  two components of, 92-94, 206, 208
Friction, 11, 61
  and activation energy, 225, 264

G

Gas constant (R), xii
Gases, perfect, *see* Perfect gases
Gibbs-Duhem equation, 205-206, 229
Gibbs free energy (G), 85, *see also* Free energy
Glass electrode, 198-199, 201-202

H

H (enthalpy), 17, 107
Heat capacity, 36-37
  of perfect gases, 70
Heat engine activity in organisms, 226, 263
Heat, high-temperature and low-temperature, 64
Heat, meaning of, 13-14
  and thermal energy, 14, 51-52
Heat of transfer (Q^*), 246
  and activation energy, 263
Helmholtz free energy (F), 85, *see also* Free energy
Hess's Law, 163

I

I (ionic strength), 179
I (intensity of radiation), 220
Ideal
  dilute solutions, 150
  solutions, 149
Ideal gases, *see* Perfect gases
Increase of Entropy, Principle of, 82
Independence of chemical reactions, 260

Inhibitors as tests for active mechanisms, 261-262, 266-267
Integration, 39
Intensity of radiation (I), 220
  specific ($\mathscr{I}$), 221
Intensive properties, 135, 136, 137
Internal energy, 16, 17, 53
Ionic activity coefficient, *see* Activity coefficient
Ionic equilibria, 5
Ionic strength (I), 179
Ions, electrochemical potential of, 175-176
Irreversibility
  chemical, 48
  thermodynamic, 41 *et seq.*, 58, 259, 263
Isoaffine reactions, 254
Isolated systems, 19
Isothermal expansion
  of perfect gases, work done in, 54-55
  of radiation, 219
  source of work in, 66, 67

K

k (Boltzmann's constant), xii
K (chemical equilibrium constant), 159-160
KCl, use of in salt bridge, 197
Kinetic energy, 13
Kirchhoff's Law, 216, 223

L

Latent heat, 246-247
Leaf as heat engine, 226
Liquid junction potentials, 197
Living systems
  criterion for spontaneity in, 109-110
  thermodynamics and, 4-6, 269-272
Le Chatelier, Principle of, 12, 43, 52, 90, 104

M

Macroscopic, 6, 14, 97, 102
Macrostate, 114
Maximum, condition for, 33
Maxwellian distribution, 3
'Meaningless, physically', 156
Mechanical analogies, 112
Membrane
  equilibria, 192-194
  potential, measurement of, 200-210
  semi-permeable, 60

Membranes, significance of, 101, 262-263
Memory, Principle of, 271
Metabolic
  pattern, efficiency of, 255-259, 263-265
  rate, importance of, 259-260
Metastable equilibrium, 98-99, 223-224
Microscopic level, 102
Microscopic systems, thermodynamics and, 115, 270
Microstate, 114
Minimum, condition for, 33
Minimum Entropy Production, Principle of, 251-254
Mitochondria, 67, 115, 263
Molality, 146
Molarity, 146
Mole, xii, 146
Molecular theory and thermodynamics, 6
Mole fraction, 146

N

Non-equilibrium, applicability of thermodynamics in, 4-6, 233-234
Non-ideality, effect of on ionic accumulation ratios, 187-188

O

Ohm's Law, 234, 236
Onsager affinity, see Affinity
Onsager cross-coefficients, see Cross coefficients
Onsager's reciprocal relations, 235-236
Open system, 129
Order of stationary state, 253
Osmotic potential, 55
  and ambient pressure, 88
  and vapour pressure, 90
Osmotic unbalance in Donnan equilibrium, 189
Oxidation-reduction (redox) potentials, 110, 199-200

P

Partial
  differential coefficients, 26 et seq.
  molar quantities, 37-38
    free energy, 136
    volume of water, values of, 211

Partition coefficients, 169-170
Passive processes, 6, 203, see also Active processes and Active transport
Pathway, meaning of, 15
Perfect gases, 49 et seq.
  entropy and volume of, 122-125
  heat capacity of, 70
  internal energy of, and temperature, 53
  work done in isothermal expansion of, 54-55
Permeability
  relation to of Onsager coefficients, 251
  temperature dependence of, 265-266
pH measurement, 198-199
Phase, 130
Phases, electrical potential between, 173-175
Phenomenological equations, 234
Phosphates, high energy, 227-228
Photosynthesis, 213 et seq.
  quantum efficiency of, 226-227
'Physically meaningless', 156
Planck's constant (h), xii
  radiation formula, 230-231
Plasmolysis, water movement in, 208
Polarization of light, 223
Potential
  chemical, see Chemical potential
  diffusion, 174
  Donnan, 183, 186-187
  electrical, 173-175
  electrochemical, 175-176
  liquid junction, 197
  membrane, 200-210
  osmotic, 55
  oxidation-reduction, 110, 199-200
  suction, 4, 206-207
Potential energy, 13
Pressure, influence of, on transport equilibrium, 155, 190-192
Principle
  of Conservation of Energy, 10 et seq.
  of Dissipation of Energy, 79
  of Equipartition of Energy, 126
  of Evolution, 271
  of Increase of Entropy, 82
  of Le Chatelier, 12, 43, 52, 90, 104
  of Memory, 271

Principle
  of Minimum Entropy Production, 251-254
  of Virtual Work, 106
Probability and entropy, 116
Process, 7
Processes, vector and scalar, 101-102

Q

Q_{10} (temperature coefficient), 265
Q^* (heat of transfer), 246
Quantum efficiency of photosynthesis, 226-227
Quantum of radiant energy, 230

R

Radiant energy, quantum of, 230
Radiation
  black-body, 217, 230-231
  change of temperature of on scattering, 221-222
  entropy of, 222-224
  intensity of (I), 220
  isothermal expansion of, 219
  Planck's formula for, 230
  reflection of, 56, 223
  temperature of, 217-222, 230-231
Rate
  of attainment of equilibrium, 5, 100-101
  of processes in organisms, 259
Rates, thermodynamics and, 3
Redox
  electrode, 199-200
  potentials, 110, 199-200
Reflection, diffuse and specular, 222-223
Relative Donnan strength, 184
Respiration, free energy change in, 165-168
Reversible electrodes, 196-197
Reversibility
  chemical, 48
  thermodynamic, 41 et seq.
Root, thermodynamics of water movement across, 208

S

S (entropy), 62 et seq., 80 et seq. See also Entropy
Salt
  respiration, 254
  uptake, 267

Salt bridge, 197-198
Salts, chemical potential of, 177-179
Scalar
  flow of heat, 261
  processes, 101
Scattering of radiation, effect on temperature, 221-222
Second Law, statements of, 60, 62, 78, 79
Semi-permeable membrane, 60
Solutions, 149-150
Soret effect, 254
Source of work in isothermal processes, 66-67
Specific intensity of radiation ($\mathscr{I}$), 221
Speed of reversible operations, 47
Stable equilibrium, 98-99
Standard
  chemical potential, see Chemical potential
  electrode, 196-199
  free energy change, 161
State, thermodynamic, 14, 63
Stationary states, 98, 251-253
Statistical implications of free energy, 160
Statistical Thermodynamics, 7
Stefan's Law of radiation, 231
Stirling's approximation, 123-124
Stoichiometrical ratio, incompatible with active processes, 258, 262
Suction potential (S.P.), 4, 206-207
  and vapour pressure, 209-210
System, 1, 14, 19, 129

T

Temperature coefficient (Q_{10}), 265
  absolute, 50, 131
  dependence of chemical potential on, 157
  dependence of free energy on, 168-169
  dependence of permeability on, 265-266
  differences of within cells, 262-264
  of radiation, 217-222, 230-231
Thermal diffusion (Soret effect), 254, 271
Thermal energy, 14, 51 et seq., 125 et seq.
  and heat, 14, 51-52
Thermodynamic
  equations, addition of, 162 et seq., 171-172
  irreversibility, 41 et seq., 58, 259, 263
  state, 14, 63

Thermodynamics
    and efficiency, 213-215
    and equilibrium, 2-3
    and living systems, 4-6, 269-272
    and mechanism, 7
    and microscopic systems, 115, 270
    and molecular theory, 6
    and non-equilibrium, 4-6, 233-234
    and rates, 3
    First Law of, 11 et seq., 19
    Second Law of, 60, 62, 78, 79
    uses of, 8
Thermo-osmosis, 239-248
Thermo-osmotic pressure, 247
Total differential, 29-31
Transpiration, 110, 208
Transport equilibrium constant, 169-170
Turgor pressure, effect of on transport equilibrium, 155, 190-192

U

U (internal energy), 16-17, 53
Uphill movement, 266, see also Active transport
Useful work, 95

V

Van't Hoff isotherm, 160
Vapour pressure
    and osmotic potential, 90
    and suction potential, 209-210
Variable of state, 17
Vector processes, 101
Velocity of light, xii
Virtual Reactions, 257–258
Virtual Work, Principle of, 106

W

Water
    active transport of, 203
    equilibrium, 203
    movement in plants, examples of 208
    partial molar volume of, 211-212
Work
    chemical, 13
    maximum, in isothermal changes, 84-85
    meaning of, 9, 13
    Principle of Virtual, 106
    useful, and Gibbs free energy, 95

X

Xylem flow, thermodynamics of, 208